FAO
TECHNICAL
GUIDELINES FOR
RESPONSIBLE
FISHERIES

5

Suppl. 6

AQUACULTURE DEVELOPMENT

6. Use of wild fishery resources for capture-based aquaculture

FOOD AND AGRICULTURE ORGANIZATION OF THE UNITED NATIONS
Rome, 2011

The designations employed and the presentation of material in this information product do not imply the expression of any opinion whatsoever on the part of the Food and Agriculture Organization of the United Nations (FAO) concerning the legal or development status of any country, territory, city or area or of its authorities, or concerning the delimitation of its frontiers or boundaries. The mention of specific companies or products of manufacturers, whether or not these have been patented, does not imply that these have been endorsed or recommended by FAO in preference to others of a similar nature that are not mentioned.

The views expressed in this information product are those of the author(s) and do not necessarily reflect the views of FAO.

ISBN 978-92-5-106974-5

All rights reserved. FAO encourages reproduction and dissemination of material in this information product. Non-commercial uses will be authorized free of charge, upon request. Reproduction for resale or other commercial purposes, including educational purposes, may incur fees. Applications for permission to reproduce or disseminate FAO copyright materials, and all queries concerning rights and licences, should be addressed by e-mail to copyright@fao.org or to the Chief, Publishing Policy and Support Branch, Office of Knowledge Exchange, Research and Extension, FAO, Viale delle Terme di Caracalla, 00153 Rome, Italy.

© FAO 2011

PREPARATION OF THIS DOCUMENT

These technical guidelines on the use of wild fishery resources for capture-based aquaculture have been prepared by the Fisheries and Aquaculture Department of the Food and Agriculture Organization of the United Nations (FAO) under the coordination of Alessandro Lovatelli, Aquaculture Officer, Fisheries and Aquaculture Resources Use and Conservation Division. The production of the guidelines has been supported by the Government of Japan through a Trust Fund Project (Towards Sustainable Aquaculture: Selected Issues and Guidelines) and by the FAO Regular Programme. This project aimed to address selected key issues of sustainability in global aquaculture practices and development.

The initial discussions in preparation of the guidelines took place at the FAO expert workshop *Technical Guidelines for the Responsible Use of Wild Fish and Fishery Resources for Capture-based Aquaculture Production*, held in Hanoi, Viet Nam, from 8 to 12 October 2007. To develop these guidelines, eleven species-specific and two general review papers were prepared. They included both marine and freshwater examples and covered ecological, socio-economic and livelihood aspects of capture-based aquaculture.

The experts who attended the workshop and contributed in discussions and with inputs to the guidelines were: Don Griffiths (Ministry of Agriculture and Rural Development, Viet Nam), Øystein Hermansen (Norwegian Institute of Fisheries and Aquaculture Research, Norway), Robert Pomeroy (University of Connecticut-Avery Point, United States of America), Anders Poulsen (Ministry of Agriculture and Rural Development, Viet Nam), Victor Pouomogne (Institute of Agricultural Research for Development, Cameroon), Patrick Prouzet (Institut français de recherche pour l'exploitation de la mer, France), Kjell Midling (Norwegian Institute of Fisheries and Aquaculture Research, Norway), Mohammed Mokhlesur Rahman (Center for Natural Resource Studies, Bangladesh), Makoto Nakada (Tokyo University of Marine Science and Technology, Japan), Francesca Ottolenghi (Halieus, Italy), Magdy Saleh (General Authority for Fish Resources Development, Egypt), Yvonne Sadovy de Mitcheson (University of Hong Kong, Hong Kong Special Administrative Region), Colin Shelley (YH & CC Shelley Pty Ltd, Australia), Choi Kwang Sik (Cheju National University, Republic of Korea), Pham An Tuan (Research Institute for Aquaculture No. 1, Viet Nam), and Mark Tupper (WorldFish Center, Malaysia).

The technical guidelines were finalized by Yvonne Sadovy de Mitcheson, with contributions and comments provided by a number of FAO fisheries and aquaculture officers, including Devin Bartley, Gabriella Bianchi, Junning Cai, Simon Funge-Smith, Mohammad Hasan, Nathanael Hishamunda, Alessandro Lovatelli, Gerd Marmulla, Doris Soto, Rohana Subasinghe, Sachiko Tsuji and Diego Valderrama.

Layout formatting was done by José Luis Castilla Civit.

FAO.
Aquaculture development. 6. Use of wild fishery resources for capture-based aquaculture.
FAO Technical Guidelines for Responsible Fisheries. No. 5, Suppl. 6.
Rome, FAO. 2011. 81 pp.

ABSTRACT

The aquaculture of commercially valuable fish and invertebrate species is growing rapidly worldwide and has become a critically important additional means of production of freshwater and seafood at a time when many natural populations are declining in the wild. Capture-based aquaculture (CBA) is defined as the practice of collecting live material from the wild and its use under aquaculture conditions. It makes a significant contribution to aquatic production and livelihood generation. It encompasses a range of activities, from the capture of larvae, juveniles and subadults of desirable fish and invertebrate species as seed material for grow-out in captive conditions, to the taking of adults as broodstock and the use of wild-caught fishes and invertebrates for feed. Because CBA combines culture activities with exploitation of natural resources, there is potential for competition and conflict among fishing sectors that target different life history phases of target species and for impacts on the environment through overfishing or habitat damage. There are very few species produced by aquaculture that have little, or no, dependence on wild populations of target and non-target species. This means that the aquaculture of many species is still reliant on the sourcing of organisms from natural populations for some part of the operation, or with impacts to the wild fisheries in some manner as a result of that activity. The management and conduct of operations that have these effects, therefore, need to take account of both fishery and aquaculture considerations and good practices.

Until recently, CBA attracted little attention as an activity distinct from hatchery-based aquaculture (HBA) for monitoring and management consideration and indeed it has typically been treated in the same way as HBA. However, while the use of wild-caught resources for feed in aquaculture facilities is similar for both CBA and HBA, the heavy dependence of CBA on wild resources for seed and its implications for wild populations have been increasingly recognized in the last decade.

The long-term goal of most forms of aquaculture is eventually to transition from CBA to fully HBA; however, there is a range of biological, socio-economic and practical reasons why this is unlikely to occur for many species, or in some cases, where this may even be undesirable or unnecessary. It must be recognized that CBA is an important and essential part of the aquaculture industry, but to ensure that its contributions lead to long-term societal and environmental benefits it must be operated sustainably and according to the FAO Code of Conduct for Responsible Fisheries and within the framework of an ecosystem approach to management. Recognizing that CBA will continue to provide important or essential inputs to aquaculture operations and that it is the starting point for the aquaculture of any species has led to the development of these technical guidelines for the responsible management and conduct of this activity.

Specifically, these guidelines address the actual and potential impacts of wild-seed harvest on target and non-target (bycatch), including threatened species, biodiversity and on the environment and marine ecosystem. The guidelines also consider capture and post-collection practices, grow-out, feed and broodstock, social and economic factors, and governance considerations. These technical guidelines identify CBA principles and guidelines for good practices and provide numerous illustrative case studies from a diverse range of species and fisheries.

CONTENTS

ABBREVIATIONS AND ACRONYMS

CBA	capture-based aquaculture
CBF	culture-based fisheries
CITES	Convention on International Trade in Endangered Species of Wild Fauna and Flora
Code	Code of Conduct for Responsible Fisheries
COFI	FAO Committee on Fisheries
EAA	ecosystem approach to aquaculture
EAF	ecosystem approach to fisheries
EIFAAC	European Inland Fisheries and Aquaculture Advisory Commission
F	fishery mortality
HBA	hatchery-based aquaculture
ICES	International Council for the Exploration of the Sea
IUCN	International Union for Conservation of Nature
IUU	illegal, unreported and unregulated (fishing)
M	natural mortality
PL	post-larvae
RFMO	regional fisheries management organization

BACKGROUND

1. From ancient times, fishing from oceans, lakes and rivers has been a major source of food, a provider of employment and other economic benefits for humanity. Ocean productivity seemed particularly unlimited. However, with increased knowledge and the dynamic development of fisheries and aquaculture, it was realized that living aquatic resources, although renewable, are not infinite and need to be properly managed if their contribution to the nutritional, economic and social well-being of the growing world's population was to be sustained.

2. However, for nearly three decades, because of the dramatic increase of pollution, abusive fishing techniques worldwide, and illegal, unreported and unregulated fishing, catches and landings have been shrinking and fish stocks declining, often at alarming rates.

3. Stock depletion has negative implications for food security and economic development and reduces social welfare in countries around the world, especially those relying on fish as their main source of animal protein and income such as subsistence fishers in developing countries. Living aquatic resources need to be properly managed if their benefits to society are to be sustainable.

4. Sustainability of societal benefits requires a recovery of depleted stocks and maintenance of the still-healthy ones through sound management. In this regard, the adoption of the United Nations Convention on the Law of the Sea, in 1982, was instrumental. The law provides a new framework for the better management of marine resources. The new legal regime of the oceans gave coastal States rights and responsibilities for the management and use of fishery resources within the areas of their national jurisdiction, which embrace some 90 percent of the world's marine fisheries.

5. In recent years, world fisheries have become dynamically developing sectors of the food industry, and many States have strived to take advantage of their new opportunities by investing in modern fishing fleets and processing factories in response to growing international demand for fish and fishery products. It became clear, however, that many fisheries resources could not sustain an often uncontrolled increase of exploitation. Overexploitation of important fish stocks, modifications of ecosystems, significant economic losses, and international conflicts on management and fish trade still threaten the long term sustainability of fisheries and the contribution of fisheries to food supply.

6. In light of this situation, while recognizing that the recovery of depleted stocks is still urgent and avoiding depleting still-healthy stocks as important, FAO Member States have expressed the need to further develop aquaculture as the only immediate way to bridge the gap between the declining capture fisheries output and the increasing world demand for seafood.

7. Indeed, in the last three decades, aquaculture has recorded a significant and most rapid growth among the food-producing sectors and has developed into a globally robust and vital industry. However, aquaculture also has been shown at times to carry the potential to cause significant environmentally and socially adverse impacts.

8. Thus, the Nineteenth Session of the FAO Committee on Fisheries (COFI), held in March 1991, recommended that new approaches to fisheries and aquaculture management embracing conservation and environmental, as well as social and economic, considerations were urgently needed. FAO was asked to develop the concept of responsible fisheries and elaborate a Code of Conduct to foster its application.

9. Subsequently, the Government of Mexico, in collaboration with FAO, organized an International Conference on Responsible Fishing in Cancún in May 1992. The Declaration of Cancún, endorsed at that Conference, was brought to the attention of the United Nations Conference on Environment and Development Summit in Rio de Janeiro, Brazil, in June 1992, which supported the preparation of a Code of Conduct for Responsible Fisheries. The FAO Technical Consultation on High Seas Fishing, held in September 1992, further recommended the elaboration of a code to address the issues regarding high seas fisheries.

10. The One Hundred and Second Session of the FAO Council, held in November 1992, discussed the elaboration of the Code, recommending that priority be given to high seas issues and requested that proposals for the Code be presented to the 1993 session of the Committee on Fisheries.

11. The Twentieth Session of COFI, held in March 1993, examined in general the proposed framework and content for such a Code, including the elaboration of guidelines, and endorsed a time frame for the further elaboration of the Code. It also requested FAO to prepare, on a "fast track" basis, as part of the Code, proposals to prevent reflagging of fishing vessels which affect conservation and management measures on the high seas. This resulted in the FAO Conference, at its Twenty-seventh Session in November 1993, adopting the Agreement to Promote Compliance with International

Conservation and Management Measures by Fishing Vessels on the High Seas, which, according to FAO Conference Resolution 15/93, forms an integral part of the Code. It was also recognized and confirmed that issues of responsible aquaculture development and aquaculture sustainability should be addressed in the formulation process so that these be appropriately covered in the envisaged Code.

12. This implicit recognition of the importance of governance in aquaculture is underlined in Article 9.1.1 of the Code, which requires states to "establish, maintain and develop an appropriate legal and administrative framework to facilitate the development of responsible aquaculture". In addition, at the beginning of the new millennium there is growing recognition of the significant potential for and implications of the use of ocean and coastal waters for mariculture expansion. The outstanding issue in this area is that, unlike in capture fisheries, the existing applicable principles of public international law and treaty provisions provide little guidance on the conduct of aquaculture operations in these waters. Yet, experts agree that most of the future aquaculture expansion will occur in the seas and oceans, certainly further offshore, perhaps even as far as the high seas. The regulatory vacuum for aquaculture in the high seas would have to be addressed should aquaculture operations expand there.

13. The Code was formulated so as to be interpreted and applied in conformity with the relevant rules of international law, as reflected in the 10 December 1982 United Nations Convention on the Law of the Sea. The Code is also in line with the Agreement for the Implementation of the Provisions of this Law, namely the 1995 Conservation and Management of Straddling Fish Stocks and Highly Migratory Fish Stocks. It is equally in line with, *inter alia*, the 1992 Declaration of Cancún and the 1992 Rio Declaration on Environment and Development, in particular Chapter 17 of Agenda 21.

14. The development of the Code was carried out by FAO in consultation and collaboration with relevant United Nations agencies and other international organizations, including non-governmental organizations.

15. The Code of Conduct consists of five introductory articles: Nature and scope; Objectives; Relationship with other international instruments; Implementation, monitoring and updating; and Special requirements of developing countries. These introductory articles are followed by an article on General principles, which precedes the six thematic articles on Fisheries management, Fishing operations, Aquaculture development, Integration of fisheries into coastal area management, Post-harvest practices and trade,

and Fisheries research. As already mentioned, the Agreement to Promote Compliance with International Conservation and Management Measures by Fishing Vessels on the High Seas forms an integral part of the Code.

16. The Code is voluntary. However, certain parts of it are based on relevant rules of international law, as reflected in the United Nations Convention on the Law of the Sea of 10 December 1982. In capture fisheries, the Code also contains provisions that may be or have already been given binding effect by means of other obligatory legal instruments among the Parties, such as the Agreement to Promote Compliance with Conservation and Management Measures by Fishing Vessels on the High Seas, 1993. In aquaculture, the provisions of the Code implicitly encourage participatory governance of the sector, which extends from industry self-regulation, to co-management of the sector by industry representatives and government regulators and to community partnerships. Compliance is self-enforced or enforced by peer pressure, with industry organizations having the ability to exclude those who do not comply and with governments only checking periodically.

17. The Twenty-eighth Session of the Conference in Resolution 4/95 adopted the Code of Conduct for Responsible Fisheries on 31 October 1995. The same Resolution requested FAO *inter alia* to elaborate appropriate technical guidelines in support of the implementation of the Code in collaboration with members and interested relevant organizations.

18. The expanding role and increasing contribution of aquaculture to economic growth, social welfare, as well as global food security was recognized and reiterated at international levels such as the 1995 FAO/Japan Conference on the Contribution of Fisheries and Aquaculture to Food Security, the 1996 World Food Summit, the 1999 Ministerial Meeting on Fisheries, the 2000 FAO/NACA (Network of Aquaculture Centres in Asia and the Pacific) Conference on Aquaculture in the Third Millennium and its Bangkok Declaration and Strategy, and most recently, the 2009 World Summit on Food Security.

19. The application of the ecosystem approach to fisheries and aquaculture as a strategy for the development of the sector contributes to the implementation of the provisions of the Code, thereby enforcing the technical, ecological, economic and social sustainability of the industry.

20. Article 7 of the Code of Conduct for Responsible Fisheries focuses on management of wild fisheries and Article 9 on aquaculture. FAO has produced a number of technical guidelines on specific issues of responsible fisheries

and aquaculture to assist Member States in the implementation of the Code. It is noteworthy that the *FAO Technical Guidelines for Responsible Fisheries No. 5 – Aquaculture Development* points out that detailed guidelines on specific issues and topics covered by Article 9 of the Code will be developed by FAO in collaboration with interested partners and identifies the need to provide specific guidelines on certain types of aquaculture systems.

21. These technical guidelines provide a framework for sustainable capture-based aquaculture (CBA) within the overall context of the FAO Code of Conduct for Responsible Fisheries. They provide general principles, guidance on evaluating the suitability of existing or proposed CBA and guidance on wild capture fisheries for CBA live material, inclusive of seed material and broodstock specimens.

22. As CBA involves both capture fishery and aquaculture components, the principles and guidance enshrined within both the ecosystem approach to fisheries and the ecosystem approach to aquaculture are highly relevant and form the foundation for these guidelines.

1. INTRODUCTION

1.1 Capture-based aquaculture

Fisheries and aquaculture have been and remain important sources of food for humanity, as well as a provider of employment and other benefits. These two activities are often considered to be very different, often compared as the difference between hunting and farming. They are different in many aspects of what is done and who it is done by. Aquaculture certainly owes its origins to fishing, where wild fish or shellfish were trapped or settled in ponds or cages and then cultured to a larger size. Indeed, such systems continue to exist today and still provide significant amounts of the global production from aquaculture. The targeted capture of seed or broodstock for aquaculture operations is a more recent development and one that can result in impacts on wild populations, their habitats and non-targeted species. Another well-known linkage between capture fisheries and aquaculture is the direct use of wild fish to feed cultured animals. Although this may be viewed solely as a fisheries management concern, unrelated to the aquaculture operation, the interdependence of the fisheries and aquaculture must be recognized in order to manage the two effectively.

Previously, these interdependencies between fisheries and some forms of aquaculture had not been widely acknowledged as a distinctive activity and were simply considered to be a form of aquaculture, unrelated to the conduct or management of capture fisheries. The form of aquaculture that is directly linked to capture fisheries operations is termed "capture-based aquaculture" (CBA) and it can be considered as the practice of collecting live material from the wild and its subsequent use in aquaculture. It is, therefore, an aquaculture operation that involves some form of wild capture fishery activity for deriving seed material, broodstock specimens or feed up to the point of sale or trade.

Because of its linkage to capture fisheries, it is now recognized that CBA can cause ecosystem effects, such as contributing to or even drive overfishing, and negatively impacting non-target species and habitats. When badly managed, such CBA can affect ecosystem functions and services with negative environmental, social and economic consequences. In the case of CBA, which includes significant wild capture, the practice can also contribute to threats to species from overfishing. In such cases, CBA is or has been pursued unsustainably, with negative impacts on wild animal resources, the environment and on some sectors of society.

It is also clear that responsible CBA can contribute positively to livelihoods and economies, as demonstrated through examples of sustainable CBA. Capture-based aquaculture is the necessary first step in the development of fully closed-cycle aquaculture. It can provide a significant supplement to the production of aquatic resources, an outcome of increasing importance given the declining capture rates in many wild stocks of fishes and invertebrates. In doing so, CBA can be a significant economic activity, providing many livelihoods and producing food in a manner that can be conducted sustainably.

The ecosystem approach to fisheries (EAF) and the ecosystem approach to aquaculture (EAA) have three main objectives: (i) ensuring human well-being; (ii) ensuring ecological well-being; and (iii) facilitating the achievement of both, i.e. effective governance of the sector/areas where aquaculture occurs and has potential for development. In these guidelines, the term "sustainability" refers to the potential for long-term maintenance of human well-being, which in turn depends on the well-being of the natural world and the responsible use of its limited resources. Sustainable CBA, therefore, demands both sustainable practices at the level of the target species, as well as taking responsibility for its interactions in the ecosystem context.

1.2 Terms and definitions

Given that there is no existing definition for CBA, that CBA is a significant activity, and that CBA is not specifically incorporated in the definition of "aquaculture" by FAO, there is a need for a concise and clear definition for use in these guidelines. A suitable starting point is the definition developed by Ottolenghi *et al.* (2004), which states: "Capture-based aquaculture is the practice of collecting 'seed' material – from early life stages to adults – from the wild, and its subsequent on-growing in captivity to marketable size, using aquaculture techniques."

While this definition makes an important contribution to advancing the understanding of CBA, it is largely focused on the grow-out phase of aquaculture. There is a need for a broader definition that can adequately incorporate the wider range of CBA activities and issues, e.g. the capture (i.e. collection) of broodstock or seed material from the wild for aquaculture use.

The following term is proposed as a definition of CBA: "Capture-based aquaculture is the practice of capturing or collecting live material from the wild and its subsequent direct use in aquaculture."

Based on this, it should be noted that CBA, in addition to the taking of seed, includes the collection of broodstock from the wild for use in hatcheries, whereby the aquaculture system requires repeated replenishment from the wild stock for each production cycle generation produced. Furthermore, the key aspect of this definition, which has not elsewhere been considered in aquaculture practices, is that there can be significant wild capture or collection involved in relation to some types of grow-out operations that have previously been considered only as "aquaculture" and unrelated to "fisheries".

One reason for the frequent confusion between capture fisheries and what constitutes aquaculture is attributable to the widely used FAO definition for aquaculture: "Aquaculture is the farming of aquatic organisms including fish, molluscs, crustaceans and aquatic plants. Farming implies some sort of intervention in the rearing process to enhance production, such as regular stocking, feeding, protection from predators, etc. Farming also implies individual or corporate ownership of the stock being cultivated. For statistical purposes, aquatic organisms which are harvested by an individual or corporate body which has owned them throughout their rearing period contribute to aquaculture while aquatic organisms which are exploitable by the public as a common property resource, with or without appropriate licences, are the harvest of fisheries."

The fact that the source of stock used for "aquaculture" grow-out in some systems may be derived from wild capture fisheries is not specified in the FAO definition of "aquaculture". Capture-based aquaculture clearly falls between the definitions of "true fisheries" and "true aquaculture", being closer to one or the other, depending upon the type of system and the degree of dependence on wild fisheries resources.

Examples of the forms of CBA that involve a component of wild capture together with a component of aquaculture are extremely varied, ranging from the "fattening" of wild caught tuna, to the catch of juvenile grouper (and many other species) for grow-out in culture cages or ponds to market size, the catch of gravid male seahorse and grow-out of the young they bear, the collection and farming of wild clam spat, or the use of brushwood aggregating devices to facilitate fish and shrimp seed capture for subsequent grow-out. These guidelines also provide other examples in text boxes to illustrate particular aspects of CBA.

The need to manage the wild capture of seed and broodstock destined for grow-out in aquaculture operations and to ensure that such harvest is conducted sustainably must be a consideration for any CBA operation, as for all activities that involve wild harvest fisheries.

It is also useful to clarify that there are other related production operations that are not considered CBA, based on the agreed definition above; CBA is not "culture-based fisheries" (as previously defined by FAO) and it is not "live storage" (as described in Appendix 1, Glossary of definitions).

1.3 Purpose, objectives and scope of the guidelines

These technical guidelines provide a framework for sustainable CBA and within the overall context of the FAO Code of Conduct for Responsible Fisheries (the Code) (FAO, 1995). They provide general principles and guidance on evaluating the suitability of existing or proposed CBA. Since it has only recently been recognized that CBA is a distinct activity, the guidelines provide numerous examples – from a wide taxonomic and geographic range – that highlight the challenges of achieving sustainable CBA, the implications of the failure to do so and how to improve existing practices.

The need for considering fisheries and aquaculture activities within a broader ecosystem context, including social, economic and governance considerations for sustainability and food security, has led to the adoption of innovative approaches such as the EAF and the EAA. Because CBA involves both capture fishery and aquaculture components, the principles and guidance enshrined within both the EAF and EAA are highly relevant and form the foundation for these technical guidelines.

The broad purpose of these guidelines is to implement an ecosystem-based approach to CBA practices in order to ensure the long-term sustainable use of all the resources involved and to minimize possible adverse impacts on the environment and local communities, as stated in Articles 6 and 9 of the Code. While there may be a tendency to move from CBA to hatchery-based aquaculture (HBA) as the life cycle of target species is closed (and becomes completely reliant on broodstock maintained in captivity), the shift is rarely fully completed because of a range of biological, practical and socio-economic reasons. It may also remain economically and technically unobtainable for many species well into the long term. Although seed production in hatcheries may have been technically and economically viable for some time, there are still major aquaculture systems that remain dependent upon the capture of wild broodstock to supply these hatcheries, either on a regular or periodic basis, or wild seed may be taken if it is cheaper or more opportune to do so. Moreover, many of the carnivorous species, even under HBA, need significant quantities of wild-caught feed. Therefore, it is highly likely that CBA practices will continue for many species to some extent.

These guidelines are intended to act as a basis for the development of evaluation criteria for the assessment of existing or proposed CBA. They may also serve as the foundation for the development of systems for monitoring or certifying such farming practices. Therefore, they are of interest to regulatory institutions, certifying bodies and producers engaged in CBA or related activities.

Capture-based aquaculture consists of two different components: capture fishery and aquaculture. Accordingly, the scope of the technical guidelines on the use of wild fish/fishery resources for CBA covers not only the issues and aspects common to capture fisheries and aquaculture, but also those that are unique to CBA. Those issues shared with capture fisheries and aquaculture are addressed in more detail by the corresponding FAO guidelines and publications, and it is recommended that users also refer to that guidance for greater detail.

Furthermore, these guidelines are not intended to address several issues related to CBA, which have been covered in other FAO technical guidelines. These include restocking, or stock enhancement, which is also referred to as culture-based fisheries (CBF) (FAO, 2008a). Diseases, health (FAO, 2007), genetic (FAO, 2008b) and feed (FAO, 2011) resource management and other aspects related specifically to the culture phase of CBA are also dealt with largely in other FAO technical guidelines, although such issues may be raised when they apply specifically to the wild capture component of CBA. However, these guidelines for CBA do apply to CBF when the source of fish or invertebrate destined for restocking is from the wild.

1.4 Structure and content of this document

In view of the diversity of CBA systems, these guidelines are not intended as detailed technical management guidelines. Rather, they are intended to highlight key principles and aspects requiring consideration for CBA development, operation and practices, and to facilitate the development of specific management approaches for individual systems. Examples that illustrate the various features of CBA practices are provided.

Following this introductory chapter, Chapter 2 reviews the application of the various global codes and agreements that relate to CBA activities, including the Code and the precautionary approaches such as the EAF and

EAA. Chapter 3 provides direction on addressing the substantive issues of CBA-related wild capture fisheries, including the management of fisheries for broodstock or seed; ecosystem and environmental impacts of CBA fisheries; use of inefficient gear and gear that results in high mortality of target broodstock/seedstock; gear that results in excessive or unacceptable bycatch; legal, regulatory and enforcement issues of CBA; animal welfare; consequences arising as a result of implementation of management measures; information needs for adaptive management; and the role of statistics in responsible CBA. It also provides direction on addressing the substantive issues of CBA activities that are unique to CBA operations after capture, including the handling, transfer and transport issues of live material; culture and grow-out issues; and feed issues.

Chapter 4 examines social and economic considerations. It acknowledges the importance of both CBA and HBA, and considers issues of livelihood, food security, conflict, gender, cultural practices, and user rights. It also focuses on both the economic advantages and the need for sufficient financial support to ensure sustainable management, and considers the economic impacts of CBA activities on other, non-CBA fisheries sectors.

Chapter 5 provides guidance on the establishment of responsible CBA-related practices inclusive of management arrangements, effectiveness and compliance, legislation, information and statistics, and education and communication/consultation with stakeholders. Also included are matters of institutional capacity for monitoring, management and enforcement and associated funding considerations. The chapter further addresses the operation of both fisheries and mariculture in the case of threatened species.

Chapter 6 provides some final thoughts on major challenges and opportunities for CBA and considers possible future developments. The final two sections are the References, which provide key reference materials for more detailed information, and the Appendixes, which include a glossary of definitions, a brief on the code of practices for alien species, and eight case studies, in addition to the 22 case studies embedded in the text of the main document, providing useful examples and information from CBA fisheries around the world and from a diverse range of fish and invertebrate species.

2. GUIDING PRINCIPLES

2.1 The FAO Code of Conduct for Responsible Fisheries

All relevant aspects of the Code and subsequent technical guidelines should be applied to the use of wild fish/fishery resources for CBA, as many of the practices, characteristics, situations and issues that are part of CBA are shared with wild capture fisheries and with aquaculture more generally. Given that CBA can involve significant impacts on wild populations, either directly through target fisheries on seed or broodstock, or indirectly through feed fisheries, bycatch or other harvesting practices, or can result in inequitable access to natural resources, sustainable CBA will typically require the application of the same set of guidelines as are relevant for wild fisheries in terms of both monitoring and management considerations and needs. Similarly, and in order to make sure that this activity is fully consistent with the Code, key issues related to sustainable aquaculture practices are also included in these guidelines.

2.2 Ecosystem approaches to capture-based aquaculture

There is broad acceptance that fishery and aquaculture activities have to be considered within the ecosystem context (including human well-being) in which they take place.

The EAF and the EAA provide frameworks to place fisheries and aquaculture activities, respectively, within the broader ecosystem context, making sure that stakeholders take full part in decision-making and in the implementation of appropriate measures and regulations. Both approaches consider people and livelihoods as an integral part of ecosystems and regard these activities as an important source of food and livelihoods. Both approaches underline the need to carry out these activities in ways that do not undermine the possibility of future generations, taking advantage of all the goods and services that aquatic ecosystems can provide. For full details, refer to FAO technical guidelines on fisheries management (FAO, 2003) and aquaculture development (FAO, 2010).

Under an EAF, any fishing activity requires the existence of a formal or informal arrangement between the fishery management authority and stakeholders, i.e. a management plan. This provides, to all those with interest in the exploited resource and the ecosystem, key information on the biology of the resource, importance to humans, and all agreed rules for managing the fishery. All possible sustainability threats that the given fishing practice may generate, as well as external threats that may affect the fishery, are considered for applying the most appropriate management action. Social and economic issues, as well as governance and institutional issues, key to the sustainability of the activity, are also considered. The management plan is an essential tool for implementing the approach. Guidance on the required steps for developing and implementing a fisheries management plan can be found in the Code and in the various technical guideline supplements.

The precautionary principle as defined in the Rio Declaration on Environment and Development, Principle 15, provides that: "In order to protect the environment, the precautionary approach shall be widely applied by States according to their capabilities. When there are threats of serious or irreversible damage, lack of full scientific certainty should not be used as a reason for postponing cost-effective measures to prevent environmental degradation" (United Nations, 1992). As the information required for decision-making is often scarce and uncertainty very high, applying the precautionary approach means that risk-averse decisions will have to be based on the best, even if incomplete, information available (see also Appendix 2).

As for the capture fisheries part, the aquaculture part of CBA should best be carried out following the principles of the EAA. The major consideration is that the negative impacts on wild fishery resources that result from the capture component of CBA do not exceed the benefits obtained from the culture component of CBA, considering both ecological and socio-economic components.

2.3 Framework for assessing sustainability of CBA

Up until the last decade, CBA was typically included within the general practice of "aquaculture" without explicitly making the link between the procurement of seed or broodstock for grow-out in CBA and the possible impacts on wild aquatic populations, and by extension, on those who depend for food and livelihood on capture fisheries of the same species. One possible reason for this is that the capture of organisms at a very early stage of development was widely assumed to have little or no impact on subsequent stock sizes. However, the need to examine these fisheries more closely has now been highlighted by the recognition that:

- very large harvest volumes, potentially exceeding sustainable levels of small seed, can be involved in relation to CBA;
- the practice of capture of juveniles and small adults for CBA in some fisheries is increasingly widespread and is often undertaken without consideration for what might be the most productive use of the stock

if managed holistically (i.e. taking into consideration total fishing pressure on all life history phases of a target species exploited by different fishery sectors);

- there is a lack of information on the relationship between capture volumes of wild seed and feed and their natural mortality rates that can be used for management; and
- there may be high levels of wasteful post-capture mortality in many CBA-related fisheries.

Given that significant CBA-based harvest of seed focuses on life history phases often not considered or taken in non-CBA fisheries, and that many capture methods have been developed specifically to harvest seed, these guidelines include consideration of relevant issues such as early natural mortality rates, gear impacts and equity of resource use, among other issues that are not elsewhere explicitly considered.

The proposed framework for assessing sustainability of CBA activities, consistent with an ecosystem approach as defined in sections 2.1 and 2.2, can be illustrated as in Figure 1.

Figure 1
A hierarchical tree framework to help identify important sustainability issues of a CBA activity

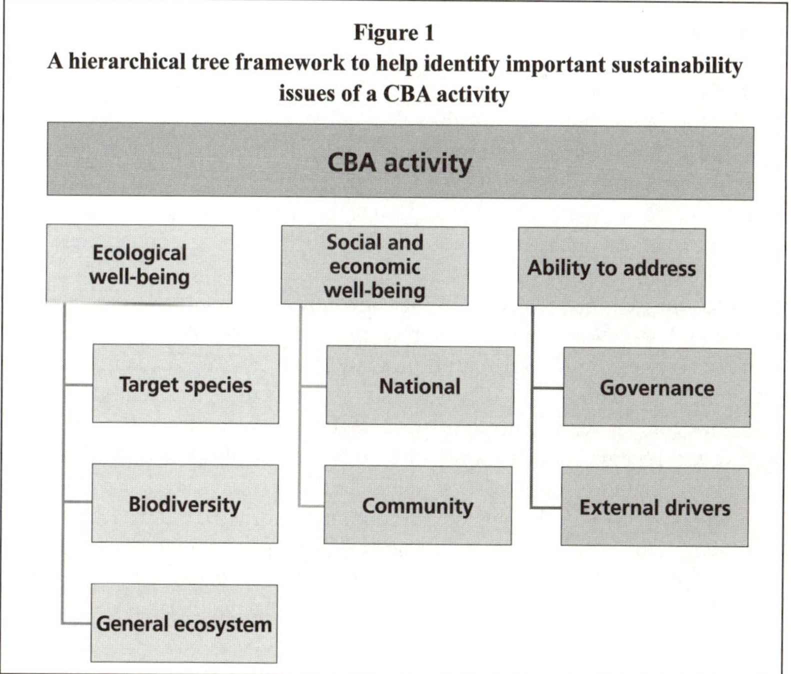

In a few cases, CBA may be conducted as part of a sustainable management plan with species that are considered to be threatened and without management, or, in more extreme cases, it may be applied for population restoration purposes. If CBA is part of a well-managed fishery, special care will need to be paid to minimize unnecessary losses resulting from poor practices and to ensure that enforcement is effective. Where HBA is feasible, this mode of production should be encouraged over CBA, following due consideration of economic and sustainability issues. However, it is possible that demand cannot be satisfied by HBA alone and, thus, that CBA will continue into the long term, such as for seahorses (Box 1). If both CBA and HBA are practised, careful documentation, and possibly tagging, will be needed for individuals of threatened species listed in Appendix II of the Convention on International Trade in Endangered Species of Wild Fauna and Flora (CITES) produced by HBA to ensure that they can be distinguished

Box 1
Global demand for dried seahorses too high for current HBA supply

Dried seahorses (*Hippocampus* spp.) are used extensively in traditional Chinese medicine and as curios, with live seahorses traded in much smaller numbers as aquarium fishes. Concerns over the unsustainable nature of the seahorse trade led to all species being listed on Appendix II of the Convention on International Trade in Endangered Species of Wild Fauna and Flora (CITES). Commercial development of seahorse aquaculture started in the 1990s, particularly in Australia, New Zealand and the United States of America, with an increasing contribution of captive-bred seahorses to the aquarium trade but not to the curio trade or to the considerably larger traditional medicine dried market, which will have to rely on wild-caught seahorses into the foreseeable future because of the numbers involved (30 million seahorses/year). While there remain technical problems with diseases and with breeding and raising some species, others are performing successfully in aquaculture. The culture of some species remains problematic over multiple generations, with reduced fecundity and developmental problems resulting in continued dependence on wild broodstock. Economic viability remains a concern in many current aquaculture operations, particularly price competition with wild-caught animals. However, the CITES listing of seahorses has increased the demand for cultured animals, and HBA production continues on a small scale.

from CBA for export purposes (see Rosser and Haywood, 2002). Cultured animals can be exported if: (i) they have been produced in captivity after at least two generations; and (ii) the latter can be demonstrated by certification or by techniques such as microchipping that identify individuals whose origins can be traced.

Animals produced through CBA may be used in restocking or recovery programmes. While such programmes are not covered by these guidelines (considered as fisheries enhancement or CBF), the conditions and practices exercised in relation to species destined for restocking or recovery programmes are the same as for other CBA operations in general. If both HBA and CBA are used to produce animals for restocking or recovery programmes, then it is advisable to seek means to assess the success rates after restocking of each mode of production for future reference and best practices.

2.4 General principles for the development, management and conduct of CBA

Capture-based aquaculture should be clearly recognized as a combination of aquaculture and capture fisheries and as such the practices and guidelines of both activities apply with recognition that CBA can have significant environmental impacts and must be managed accordingly

- Where CBA is highly dependent on wild-caught live material, i.e. when culture operations are not possible by HBA, or HBA is possible and CBA continues, management of the CBA-fishery is required and regulatory actions should be undertaken.
- Where appropriate and where they exist, regional fisheries management organizations (RFMOs) or other arrangements, in addition to measures at the national level, should ensure that CBA fishery activities are managed and monitored effectively. In the absence of an RFMO, relevant regional intergovernmental organizations should assume due responsibility.
- The ecosystem approach to fisheries and aquaculture needs to be considered and applied. This includes considerations on the type and quantities of feed used during the culture phase, seed captured for grow-out, the impacts of fishing methods and culture operations on the environment and on non-target species, and genetic issues.
- Capture-based aquaculture should be managed holistically and with due consideration paid to other fishing sectors targeting the same stock in a way that ensures the sum of fishing (and related mortalities such as during capture and transfer) does not exceed the natural mortality

of the exploited stock while taking into consideration all life history phases targeted.

- Where natural-mortality-curve information is unavailable for a new CBA fishery, ideally, no CBA activities should be undertaken for that species, except for controlled collection of live material to produce a natural mortality curve for the species and other relevant biological and socio-economic information. Alternatively, exploratory fishing could be conducted at low and controlled levels of fishing intensity, and the CBA fishery should only proceed under a set of guidelines that integrate the adaptive management concept. In all cases, new CBA activities should apply the precautionary principle and consider potential risks.
- It is important to consider the overall benefits of CBA compared to other uses of the resources. For example, if survival rates of seed following capture and during transfer and grow-out are unacceptably low, then the net benefit of such removals for CBA may be minimal and efforts should be made to utilize the wild resources in a more beneficial and sustainable way.
- Broodstock capture should be kept to a minimum and carefully monitored, especially in the case of threatened species.
- Migration routes, spawning sites and important nursery and settlement sites of CBA species should be identified, protected and managed by appropriate spatial, temporal and technical means.
- Appropriate handling methods should be applied to seeds or broodstock to minimize mortalities during transfer or grow-out.
- Holistic management will require additional controls beyond fishery management measures, such as controls on the aquaculture component of the operation. These might include licensing of hatcheries or culture operations, requirements for reporting and monitoring, regulations on quantities and size of wild seed or broodstock used.
- Monitoring and reporting of CBA fisheries should include information on the transfer into aquaculture operations (i.e. including mortalities from capture and during transfer) and, where possible, data from the aquaculture operation, such as mortality levels during the culture period.
- The effort in a CBA fishery should be monitored to enable evaluation of whether reduction or other control of fishing effort is required as part of a process of adaptive management, and what impacts such measures might have on different fishing sectors targeting the same resource.
- Equity issues need to be considered. For example, how do fishers targeting different life history phases of the same population affect one another, and can actual or potential conflicts be adequately addressed?

- All stakeholders, inclusive of fishers from all fishing sectors, fishery managers and aquaculture operators, should communicate to ensure that the linkage between the sum of capture pressure and supply and demand for seed is appropriately measured and controlled, and to ensure consultation across fishing sectors and interests.
- Where a wild capture live material or broodstock fishery is not under management and overexploitation of the wild stock and adult fishery is likely, the fishery should be halted or restricted until sustainability can be demonstrated.
- When management measures are proposed, the social and economic impacts of the management should be identified, along with mitigation measures and appropriate agencies, non-governmental organizations, international non-governmental organizations, RFMOs, etc. Means to implement programmes to mitigate social and economic impacts should also be identified.
- Countries with CBA activities should collect separate statistics on CBA with data clearly disaggregated between wild fisheries capture for CBA and culture aquaculture production.
- Animal welfare considerations must be addressed in relation to CBA operations.
- Capture-based aquaculture live material or broodstock fisheries should not come from illegal, unreported and unregulated (IUU) fishing.

3. GUIDANCE ON SUSTAINABLE CBA PRACTICES

3.1 Guidance on issues regarding ecological impacts of capture of wild live material for CBA

Capture-based aquaculture is being undertaken for many species in many countries, and new CBA developments are under way or being proposed. Therefore, there is an urgent need to evaluate the suitability of existing or proposed CBA developments in terms of ecological and biological sustainability, economic efficiency, equity, societal impact and animal welfare. The sustainable practice of CBA requires the integration of factors relevant to both sound aquaculture and sustainable capture fisheries practices, and consideration of the implications of CBA on non-CBA fishing sectors, as well as on the ecosystem. For fisheries that affect the species being exploited for CBA, this means that a management plan is required that considers the combined impacts of the fisheries on target species, as well as on the ecosystem within which the species occurs. Furthermore, the level of detail in the management plan may need to be linked to the scale and potential impact of the CBA operation. Cochrane and Garcia (2009) provide additional information on biological and other considerations relevant to managing capture fisheries in general.

A major feature of these guidelines is the need to apply the precautionary principle (see Section 2.2) in the development and operation of CBA activities. This means that if suitable measures and practices to ensure sustainable and responsible operations are not planned or in place, then: (i) the existing CBA-related fishery should be temporarily suspended until they are in place; and (ii) the proposed CBA-related fishery should not be initiated until such measures or practices are in place.

This section addresses the various issues to be considered in relation to the capture of target species for seed and broodstock for CBA, inclusive of considerations regarding spawning, nursery and settlement, natural mortality, population dynamics, life cycle, transboundary movements and stock origin. Appropriate consideration of such aspects of the biology of the target species is essential for ensuring that wild populations are managed to persist in the long term and continue to supply economic and societal benefits.

3.1.1 Impacts of CBA fisheries on wild populations of fishery resources

3.1.1.1 Spawning considerations

Maintenance of adequate spawning biomass is an essential component of the management in any fishery to avoid recruitment overfishing and consequent stock declines, and CBA is no exception. Wild populations have limits on their ability to replace individuals lost to fishing. These limits are determined by the species, the current size of the population in relation to its unfished state and by the environment in which it occurs and is exploited. It is vital to ensure that the population maintains sufficient reproductively mature adults, or spawners, also referred to as spawning biomass (refer to Section 1.3.1 of the FAO technical guidelines on Fisheries Management No. 4 [FAO, 1997]).

Aquatic organisms display a diverse range of mating strategies and tactics and some may be particularly dependent on certain habitats, seasons or conditions for successful spawning. As many CBA species are difficult to breed in captivity (and therefore, not yet qualified as HBA species), they may be especially vulnerable to overexploitation at the time or place of spawning and therefore, need special management attention. Examples include those species that form large groupings or aggregations of spawning adults that are highly predictable both temporally and/or spatially, and species that depend on specific habitats or conditions for spawning (Box 2). It should be noted that a number of aggregating species are the basis of early post-settlement-stage CBA and that aggregations could be good potential sources of ripe, high-quality broodstock.

Given the vulnerability of spawning aggregations and the fundamental need to protect sufficient spawning biomass for stock maintenance, it is essential that any activity targeting these aggregations is adequately managed as part of an overall management plan for the given population and that, if no management is in place, the fishery is suspended or allowed to proceed until management is in place.

3.1.1.2 Nursery and settlement considerations

Some species are highly dependent on specific nursery and settlement grounds, and, once known, these may become the target of fishing. If fishing pressure is too high at the settlement life-history phase, and too many organisms are removed too quickly, or settlement or nursery habitat is damaged, this could ultimately compromise the sustainability of the stock. Examples include inshore areas that are often important settlement grounds, such as estuaries, mangroves and seagrass beds.

Box 2
Spawning aggregations – the basis of some CBA fisheries

Many fish species form temporally and spatially limited groups or aggregations to spawn as their only means of annual reproduction. The high numbers of eggs produced at such times and places can form the basis of many capture-based aquaculture (CBA) fisheries if associated with heavy settlement pulses of post-larvae. Adults assembled in these spawning aggregations are often the focus of fishing and can be very rapidly depleted, indirectly affecting later settlement pulses. The majority of recorded spawning aggregations of coral reef fishes are not managed and many have been depleted by fishing. Ensuring that spawning aggregations persist and are properly managed is important for continuing fisheries, including those for CBA seed and/or broodstock of such species. Examples include the green grouper, *Epinephelus coioides*, and tiger grouper, *Epinephelus fuscoguttatus* (Serranidae), several species of rabbitfishes (Siganidae), mullets (Mugilidae), and milkfish (Channidae) (see the Web site of the Society for the Conservation of Reef Fish Aggregations at www.scrfa.org).

Important considerations include the need to adequately identify nursery/settlement areas by appropriate spatial, temporal or technical means. In some cases, threats to nursery and settlement areas may come from other human activities (e.g. cutting of mangroves, coastal development, pollution from land-based activities), and these impacts also need to be identified and addressed as appropriate.

3.1.1.3 Migration considerations

Certain species undertake upstream and/or downstream migrations as juveniles or adults during part of their life cycle to fulfil their specific biological requirements. Migrations may be to and from the sea, along shorelines, to and from floodplains, and even vertical migrations within the water column. Whether such migrations are extensive (e.g. long-distance anadromous or catadromous migrations) or short-distance, they are an important part of the life history. The period of migration can represent an important risk, especially if there are large congregations of fish that become a target of unmanaged fishing during this time or if key migration areas are disturbed or damaged (see also Section 4.3.8 of the FAO technical guidelines on Inland Fisheries No. 6, Suppl. 1 [FAO, 2008a]). As examples, annual migrations of rabbitfish (*Siganus* spp.) in Palau and elsewhere have been heavily fished to the extent that spawning runs have been severely reduced, while movements of the European eel have been severely affected by dams and weirs (Box 3). Management needs to ensure that such life history phases are not affected to the extent that they compromise persistence of the population.

Box 3
Need for management of free migration – the European eel

The European eel, *Anguilla anguilla*, migrates between oceanic and continental waters. Since the beginning of the twentieth century, numerous dams and weirs built on many rivers have impeded glass eel from migrating through estuaries and the lower parts of rivers, and yellow eel from reaching the medium and upper parts of water catchments. Demand for eel seed for aquaculture is extremely high globally and far exceeds supply. In both Europe and other regions, this has led to the development of targeted fisheries, as eel cannot be bred in hatcheries. As a result of the development of hydroelectric power stations, a large percentage of the original area suitable for eel has been lost to eel production. Although many of the obstacles in rivers are now equipped with eel passes allowing part of the eel population to migrate upstream, thereby improving the production of silver eel, mortality can still be very high when these fish migrate downstream through the hydroturbines towards the Atlantic Ocean for spawning.

If migration is a critical part of the life cycle of the target species or stock, either as part of the early developmental phase or as part of annual spawning migrations, it is essential that migration routes are adequately protected or managed by appropriate spatial, temporal or technical means.

3.1.1.4 Natural and fishing mortality considerations

Natural mortality (M) is usually very high in the early stages of the life cycle of most fish and shellfish species, and usually decreases rapidly with growth (Box 4; Figure 2). Natural mortality is an important parameter in fishery management because the relative importance of fishing mortality (F) versus M is a major consideration for sustainability. For example, where F exceeds M, the fishery is at risk of becoming overfished. If F greatly exceeds M, fishing sustainability is put at high risk and strong evolutionary selective force may take place, with unknown consequences for fished populations in the long term.

Box 4
Natural mortality and its relevance to CBA

In most exploited marine species with a planktonic larval phase, natural mortality (M) following post-settlement declines rapidly as the young organisms find suitable shelter, shift their feeding habits and learn to survive in their juvenile/adult environment. While the estimation of M in natural populations continues to be a challenge, work on fish species to date indicates that M drops to low levels within the first few weeks or months post-settlement. This means that shortly post-settlement, young organisms are reasonably expected to have a good chance of surviving into adulthood to reproduce. Examples of large juvenile and young adult fish commonly used in capture-based aquaculture (CBA), i.e. captured when M is much reduced compared with settlement levels, include tuna, grouper and humphead wrasse. As many such fish taken for CBA in these "juvenile fisheries" are likely to contribute to the next generation, their fisheries need to be managed accordingly, taking into account the numbers of both juveniles and adults captured in all of the fishing sectors to which they are exposed. Moreover, if mortality under CBA culture (i.e. grow-out) conditions is high, there may be little net overall benefit to production of removing fish from the wild before sexual maturation. Similarly, although M is high for early post-settlement organisms, their numbers are certainly not infinite and, if too many are removed for CBA, future recruitment may be significantly reduced because too few fish are surviving to adulthood.

Source: Cochrane and Garcia, 2009.

In the case of CBA fisheries, which tend to focus on smaller and younger fish or invertebrates compared with conventional capture fisheries of mature-sized fish, the implicit assumption seems to have been that M will inevitably greatly exceed F, and hence, that the level of F poses no risk to sustainability. As a result, natural mortality has largely been ignored in CBA fisheries. However, given the increasing interest in, and species diversity associated with, CBA, the massive numbers of organisms sometimes removed for CBA and the wide range of their ages post-settlement mean that both F and M are important parameters to assess, as is widely done in fisheries science and for fishery management models in general. Ideally, M at the stage of capture should be determined relative to F to ensure that F does not exceed M on a sustained basis, or substantially. In addition to F due to CBA (inclusive of capture and wastage associated with capture), it is quite possible that the same stock is

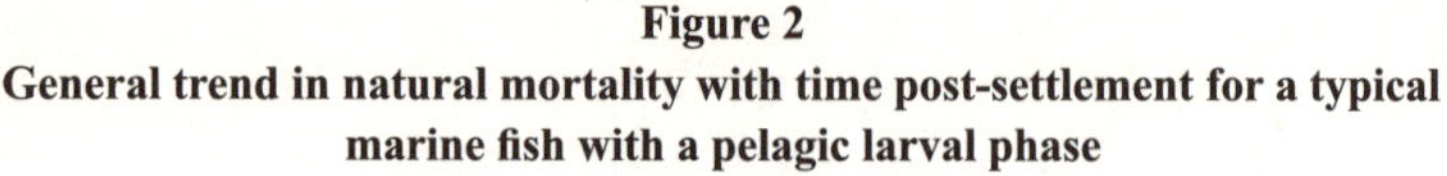

Figure 2
General trend in natural mortality with time post-settlement for a typical marine fish with a pelagic larval phase

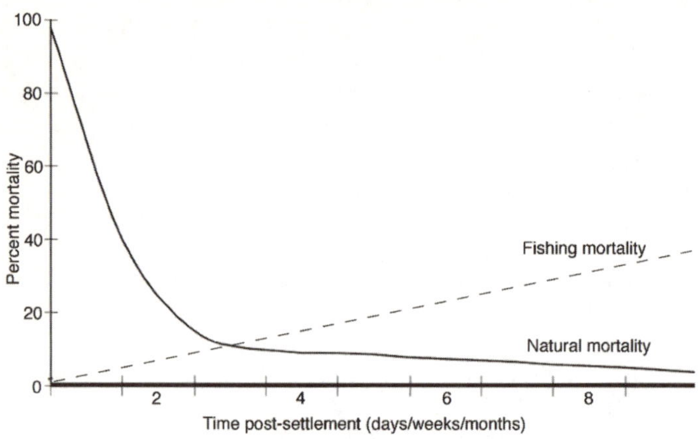

Note: The decline in percentage mortality over time will vary according to the species, and typically reaches the relatively low levels associated with adulthood quite quickly. The overall impact on the population of fishing mortality will depend on the stage (i.e. time post-settlement) that the target species is removed post-settlement, with a higher impact more likely but not limited to older individuals. The position of the fishing mortality line relative to that for natural mortality will vary somewhat according to the species and the fishery.

Source: Adapted from Sadovy de Mitcheson (2009).

being subject to additional F from other fishing sectors. If so, this would also have to be considered in management, clearly taking into account, among other factors, the overall fecundity of the target CBA species.

Key considerations related to natural and fishing mortality include the need for assessing the impacts of CBA as part of an overall assessment of the stock that considers all the sources of fishing mortality and how these compare with the levels of natural mortality of the given stock/species.

Responsible fishing should not allow more of the resource to be harvested on average and over the long term than can be replaced by the net growth of the stock. This usually involves management to maintain stock abundance in excess of some predetermined reference point which signals possible stock collapse. Failure to manage it in this way runs the risk that the resource will decline over time, leading to lower than optimal average yields and economic returns and, in extreme cases, to stock collapse. Management of fisheries requires information to be collected on trends over time in landings, sizes of organisms captured, and, if possible, on other parameters to assist management decisions, determine the relationship of the CBA-related sector to the capture fishery of the species as a whole (i.e. there may be fisheries on other sectors of the fishery), and the establishment of suitable reference points for management. Should there be both capture fisheries for direct use and for use in CBA for a given stock, assessments on the state of the resource should take into account the sum of all impacts on the population.

3.1.1.5 Transboundary considerations

For transboundary fish and invertebrate stocks, straddling fish stocks, highly migratory stocks and high seas fish stocks, where these are exploited by two or more States, the States concerned, including the relevant coastal States in the case of straddling and highly migratory stocks, should cooperate to ensure effective conservation and management of shared resource or resources. This means that management must be conducted across all relevant boundaries and beyond the spatial scale of a single catchment if the species is in export trade. The transboundary scale and complexity of management is challenging and requires considerable coordination among the States that exploit the resource. This should be achieved, where appropriate, through the establishment of a bilateral, subregional or regional fisheries organization or arrangement as encouraged by the Code. Similarly, for CBA species of interest that include transboundary stocks (e.g. eels and tuna, some migratory riverine species), CBA-related fishing mortality will have to be integrated into an existing or future transboundary management arrangement.

3.1.1.6 Species and/or stock origin considerations

The growth in CBA, its importance for aquaculture of a wide range of species and the difficulties in some areas of procuring seed have resulted in considerable volumes of international transfer of seed, often to areas far outside of the natural geographic range of the species (e.g. European eel). Given that such transfers could act as risk factors in disease transfer or introduction of exotics, with possible undesirable consequences, CBA practices need to be considered in relation to disease transfer and environmental impacts including on species diversity. Although some of these issues are also relevant to HBA, there are certain considerations specifically or indirectly pertinent to CBA practices because the impacts on biodiversity may be negative (Beveridge, Ross and Kelly, 1994).

There is a need to ensure that risk management procedures are in place to minimize the risk of disease or release of inappropriate genetic stock. In some cases where there has been extensive overfishing, translocation of closely related stock for farming purposes may be worth considering, but only once the appropriate oversight and regulatory mechanisms are in place and will be effective to minimize potential risks (see also FAO technical guidelines on Genetic Resource Management [FAO, 2008b]).

Important considerations in this respect include the need for measures, such as quarantine, risk assessment, vaccination and/or regular inspection to be applied in those situations where live material (seed or broodstock) are introduced into a region beyond the natural range of the given species.

3.1.1.7 Transition from CBA to HBA

Although the long-term goal of most CBA is that it would eventually transition to HBA to the greatest extent possible, this may not occur rapidly for many species (e.g. European eel), and in some cases it may not be necessary or desirable. Capture-based aquaculture is typically an inevitable first step in the move towards HBA, allowing much to be learned about the grow-out phase while the more challenging hatchery production is being worked out. However, based on experience to date with a wide range of species, it appears that even when HBA is achieved and reaches commercial production levels it may not fully substitute CBA for a range of biological, social and economic reasons; biological because of the need to maintain genetic diversity (which may call for continued, if much lower, dependence on wild seed and broodstock), and economic because when HBA productivity is low, it may be cheaper to source seed from the wild (e.g. yellowtail in Japan). Moreover, in some cases, many wild capture fisheries for CBA support livelihoods that are unlikely to be sustained under HBA. Thus, it seems likely that, for many species with which CBA is currently practised, economic and practical considerations will mean that some level of CBA will persist long into the future and in such cases CBA practices need to be developed or undertaken responsibly and will require to be managed accordingly (Box 5; see also Boxes A3.1, A3.2 and A3.3 for case studies on grouper and sturgeon, carp and yellowtail, respectively, in Appendix 3).

3.1.2 Impacts of CBA fisheries on biodiversity, the environment and ecosystems

The impacts of capturing wild seed for CBA can extend beyond those on the target species of interest under certain circumstances. For example, if the

Box 5
Transition from CBA to HBA – a success story in Viet Nam

The catfish *Pangasianodon hypophthalmus*, commonly known as "tra", was first artificially propagated in Thailand in 1959, but not until 1996 were pangasiid catfishes produced by hatcheries in Viet Nam. In early 2000, much of the aquaculture industry for "tra" was based on the wild capture of fingerlings. Subsequently, having successfully mastered the artificial spawning of *P. hypophthalmus*, larvae and fry/fingerling production in the Mekong Delta of Viet Nam increased substantially, and by 2008 an estimated 93 hatcheries were producing 52 billion fry for 1.3 million tonnes of fish produced and 100 000 people employed. While wild broodstock are still used seasonally to produce local seed, there are many small-scale hatcheries and nurseries supplying pangasiid seed. Indeed, more than sufficient is produced for local culture, with excess river catfish larvae and fingerlings being exported to Cambodia.

fishing gear used to take the target species damages the substrate, or takes large volumes of bycatch, its deployment needs appropriate management (Box 6). Furthermore, in some circumstances, the possible ecosystem implications of the release of gametes into non-native areas from open-water CBA systems may have to be considered.

3.1.2.1 General biodiversity, environmental and ecosystem impact considerations

The use of fisheries for live material, inclusive of seed and broodstock, can adversely affect biodiversity, the environment and the ecosystem where the fisheries occur. This can include impacts on biodiversity, from the use of chemicals, to the destruction of coral, mangrove removal, and damage to the habitat from bottom trawling gear. Ecosystem-level impacts can result from disruptions to trophic interactions or population dynamics from excessive removals or damage. While attention is paid to those activities specifically associated with CBA, issues that are more widely relevant to aquaculture production are also addressed. Article 6.6 of the Code applies: "Selective and environmentally safe fishing gear and practices should be further developed and applied, to the extent practicable, in order to maintain biodiversity and to conserve the population structure and aquatic ecosystems and protect fish quality. Where proper selective and environmentally safe fishing gear and practices exist, they should be recognized and accorded a priority in

Box 6
Bycatch of non-target species in grouper seed collection

Grouper seed for capture-based aquaculture (CBA) is collected using a wide variety of fishing methods, depending on location, size of target fish and local practices. Some methods are selective and some unselective and many of the latter generate unwanted bycatch comprising the young of non-target species. For example, the push net or scoop net, widely used in Southeast Asia for seed collection, trap both target and large numbers of non-target organisms, as well as cause damage to the substrate disturbed by the net frame. Subsequent sorting of the catch also results in the discarding of non-target species, which are typically in juvenile stages. Furthermore, an evaluation of capture rates, sizes of target species, and impact on environment and biodiversity of "gangos", a capture method for grouper seed collection used in Southeast Asia, indicated that only 1.4 percent of the total fish catch comprised target groupers and snappers while the non-target organisms typically consisted of food fish and shrimp species still too small for human consumption; most discarded dead or moribund. More careful handling could avoid the high wastage of bycatch, and additional attention is needed, as well as ways to reduce the take of, or damage to, bycatch. Seed fisheries conducted in this way can severely undermine the advantages of CBA.

Source: Mous *et al.*, 2006.

establishing conservation and management measures for fisheries. States and users of aquatic ecosystems should minimize waste, catch of non-target species, both fish and non-fish species, and impacts on associated or dependent species" (see also FAO International Guidelines on Bycatch Management and Reduction of Discards [FAO, 2011a, 2011b]).

Capture-based aquaculture should aim to reduce and, where possible, eliminate, dependence on activities and equipment that cause severe environmental impacts, e.g. excessive damage to the substrate by push nets or removal of mangroves to construct artificial reefs that attract small fish seeking shelter. This is particularly important in seed fisheries where nursery habitats may be specifically targeted. The potential for localized impact on habitats from broodstock and seedstock fisheries is high, so there is a need for responsible collection tools and methods supporting a well-managed broodstock or seedstock fishery that will ensure no significant damage to habitats.

Capture-based aquaculture is usually based on animals that have yet to reach sexual maturity, but it sometimes involves the targeting of broodstock (adults) for use in hatcheries. Collection of wild broodstock or seed can result in significant quantities of non-target species and, sometimes, in high mortalities of target seed stock. This may negatively affect recruitment of wild stock and have impacts on food webs and other ecosystem linkages. The selection of large individuals of a species for broodstock, if this involves substantial numbers of organisms being fished on a regular basis, could affect the reproductive capacity of the target stock. The absolute dependence of farming *Penaeus monodon* on wild caught broodstock resulted in very heavy selective fishing for large broodstock-sized animals, although population effects have been difficult to demonstrate because of other possible factors such as escapees, stocking of fisheries and disease.

Overall, CBA most typically involves harvesting of the target animal for culture prior to sexual maturation. However, should such animals mature and ripen in captivity, there is a possibility that they will release fertilized eggs. Generally, the conditions or location in which they are held will probably mean that these eggs will not contribute significantly or even at all to fisheries recruitment because they are not the natural spawning conditions, although some shellfish can regularly release gametes that are viable and can colonize areas in which they settle. The range of implications of cage-held organisms successfully spawning and the spawn subsequently recruiting into the wild needs to be discussed. It is an area that has been little studied and rarely considered (Box 7).

Capture-based aquaculture fisheries must consider broader ecosystem and life history issues beyond the target species and take biodiversity considerations into account during development and at all operational stages, and in particular:

- Management of CBA fisheries for live material should include measures that cover essential habitats of all stages of life cycle of the target species (by input controls, spatial protection, etc.).
- The harvest for CBA live material should not result in ecosystem overfishing through significant effects on non-target species or habitat.
- The implications of the release of fertilized eggs by ripe live material held outside of its natural range on wild stocks should be considered, especially when exotic species are involved. While unlikely to contribute significantly to natural production, there is potential for problems caused by interbreeding with wild stocks. Hence, species should not be moved out of their native areas if gamete release is likely and could lead to the introduction of foreign genetic (eggs or seed) material.

- Management of CBA fisheries should ensure maintenance of sufficient genetic diversity in wild broodstock.

Box 7
Spawning in captivity by wild-caught seed – possible implications

In some cases, as for bluefin tuna, Atlantic cod or the reef-associated humphead wrasse, capture-based aquaculture (CBA) operations may include individuals that have already attained sexual maturity. These may be destined to be broodstock or retained past sexual maturation because of market requirements (seasonal or size based). In many such cases, the adult-sized fish and invertebrates may be kept in captivity in cages in open water, and there is a possibility that fertilized eggs escape into the ocean, with both positive and negative implications. On the positive side, eggs released from adult captive animals that might otherwise be contributing to the wild population if the animals were free may not be entirely lost if they contribute to population regeneration. However, this has not been demonstrated to occur, and seems unlikely to be substantial if animals are not spawning in suitable habitats for egg survival and dispersal. As one example, those species that migrate to spawning sites and spawn only in specific areas at specific times and in specific social groupings may have less reproductive success, or produce fewer viable eggs, if spawning in captivity in other times or places. Studies on cod, *Gadus morhua*, in Norway show evidence of both non-contribution and contribution of released eggs to wild populations. On the negative side, if fish are held in captivity far away from where they were caught, the successful release of eggs could cause the introduction of exotic genetic material with uncertain consequences. While there is little evidence to date that eggs released into the wild fortuitously by ripe animals in grow-out operations survive and populate the surrounding waters, very little work has been conducted to test for this possibility. Most species have specific spawning needs in terms of locations, conditions and timing of spawning that are unlikely to be present in artificial holding stations. As one example, many groupers spawn in aggregations after migrating tens to hundreds of kilometres to traditional sites each year for a few weeks to spawn. While the evolutionary significance of these spawning sites and times is not known, they are presumably adaptive in some way, providing appropriate conditions for adults and/or eggs and larval survival. Cage conditions, at best, will be comparably suboptimal.

3.1.2.2 Environmental impacts of fishing gear and methods

As with many fisheries, bycatch and discards are problems in CBA fisheries for live seed material or broodstock, inclusive of excessive mortality. This results from suboptimal or socially unacceptable harvesting gear and methods, and from suboptimal and inappropriate treatment of bycatch. The bycatch resulting from CBA fisheries for live material often consists of large numbers of small-sized individuals of many different species (termed growth overfishing). Excessive bycatch can negatively impact biodiversity and ecosystem function(s) and the market or food potential of bycatch, should it be left to develop in the wild, is often not considered in evaluation of impacts of broodstock fishing. This bycatch also creates information problems for management, as the bycatch is often not recorded or may even be discarded. Some gear result in excessive take of bycatch partially from the way in which the gear is used or constructed. Fishing gear and methods that are known to be destructive to the environment, or can result in high mortalities of target and/or non-target species (e.g. fyke and scoop nets and poisoning) should be banned and alternatives developed (see Box 6).

Harvesting and holding of live material for CBA should be managed to limit its impact on biodiversity, the physical environment and the ecosystem, and operated in such a way that minimizes excessive mortality, bycatch or discards or produces outcomes that are otherwise socially unacceptable. Furthermore:

- Research and development should be encouraged to improve the type of gear and methods used to catch CBA species so that they are more selective and have less impact on the environment or on non-target species, or to develop new types of gear and methods that minimize bycatch and mortalities.
- Dependence on non-selective gear for CBA species should be reduced through development, promotion and training in the use of alternative types of gear.
- In cases of excessive bycatch and where no alternative gear or fishing methods are available, fishing effort for CBA species should be reduced or eliminated.
- Discarding of bycatch should be discouraged and methods to keep bycatch alive should be promoted.
- Live bycatch taken in CBA fisheries should either be released back to nature or used in CBA to reduce waste of marine biomass.
- Bycatch and discards should be documented and reported for management purposes.

- Alternative livelihood options should be explored for CBA fisheries in which existing destructive gear cannot reasonably be replaced by non-destructive (or less destructive) gear.
- Knowledge and best practices regarding responsible gear and methods in CBA fisheries should be developed and applied.

3.2 Guidance on post-collection of seed and aquaculture components for existing or proposed CBA

For cases where a decision has been made that an existing or proposed CBA activity is suitable, the following section provides guidance on ensuring that the aquaculture operations are responsible once the wild material has been collected for CBA. The general issues of handling and transport of live materials, culture and grow-out, documentation and use of fish feed are covered in this section. While such post-fishing issues are also of relevance to HBA and have been variously covered in other guidelines, those of particular or unique importance to CBA are highlighted. In addition to identifying any key problems for CBA, emphasis is also placed on the need to mitigate negative impacts of CBA to improve practices and standards of CBA operations and outcomes. Issues of relevance to the culture phase of both CBA and HBA but not specifically and directly of relevance to the wild capture component of CBA are not covered in this document. Excellent coverage of important issues ranging from best management practices at the farm level, wastes from farming, genetic considerations, health considerations, site selection and water conservation may be found in a range of FAO technical guidelines (see References) including those, among others, on health management for responsible movement of live aquatic organisms (FAO, 2007) and on genetic resource management (FAO, 2008b).

3.2.1 Live material handling and transport issues

3.2.1.1 Mortality, health and welfare of CBA live material during handling and transport

Following the capture of the live material, there are several important issues concerning the mortality, health and welfare of seed and broodstock as they move from capture to culture. There can be significant levels of mortality during the initial capture and post-capture grading and sorting process, during transport from the capture area to the aquaculture facility, and as part of the recovery or acclimatization process upon arrival in aquaculture facilities. In addition to the obvious impacts on animal health and welfare, significant mortality at this stage has negative impacts on the economic viability of CBA operations. High levels of harvest, post-harvest or pre-aquaculture mortality stimulate increased fishing effort to replace the live material lost, resulting in increased environmental impacts and a wasteful use of natural resources. Data on mortality rates are essential for managing and improving survivorship of live material from capture to aquaculture.

Even with adequate survival rates of target animals, optimal health and welfare may be compromised through sublethal stress, injury, competitive interactions of animals and other factors. Less than optimal health and welfare may reduce the ability of the live material to survive in confined farming facilities or compromise its potential for attaining optimal growth or quality during captivity, with economic implications for the CBA enterprise.

Best practices to reduce mortality and optimize the health and welfare of target and non-target animals will vary among different species and among the life stages of the live material. Best practices may include a range of issues, techniques, methods, equipment and approaches. For example, appropriate handling, transportation facilities and procedures will need to consider aspects such as tank size, sorting, conditioning, transfer practices, recovery time and inspection (Box 8). Water quality will need to be maintained to optimize

Box 8
Improved capture and handling methods to reduce mortality in the Norwegian cod CBA fishery

The survival of cod, *Gadus morhua,* during and after catch is a key consideration for its economically sustainable capture-based aquaculture (CBA). Several types of fishing gear are used in this fishery in Norway. Most of the catch is landed by medium-sized and large coastal vessels using the Danish seine, which has been modified over time to increase survival of the catch and to avoid large amounts of saithe bycatch. Some vessels vacuum pump the fish from the codend to the vessel, while others use a canvas lining inside the codend to reduce pressure on the cod during hauling. Mortality during transfer to and within the holding tank initially caused problems. In the holding tank, for example, swim bladder rupture occurred close to the surface, making the cod negatively buoyant in addition to becoming exhausted, resulting in fish piling up at the bottom of the flow-through tanks. Survivorship was increased by introducing upwelling tanks whereby the water supply comes from inlets on the bottom of the tank. The mortality associated with transfer to the holding cage was reduced by developing a cage with a rigid, flat bottom where the fish could spread out to recover physiologically and regain neutral buoyancy.

health and welfare of the live material, with appropriate monitoring and water exchange. The stocking density will need to be kept within appropriate limits for the species and the size and condition of the animals based on estimated biomass and volume. The transfer of best practices and lessons learned in reducing mortality and optimizing health and welfare is important to improve CBA overall.

The handling of CBA live material, inclusive of seed and broodstock, should aim to minimize mortality, address animal welfare issues, and make the best possible use, and incur the least wastage, of natural resources. Furthermore:

- Best practices should be identified and implemented to reduce mortality and optimize the health and welfare of the live material at all stages, from capture to captivity.
- Procedures should be documented to facilitate monitoring and reporting.
- Data on fishing, post-collection and pre-aquaculture mortality rates should be collected, compiled and analysed to identify and correct problems affecting the survivability, health and welfare of the live material.

3.2.1.2 Pharmaceutical treatments of CBA live material during handling and transport

The stress from capture and during transportation to the culture facilities affects the health and welfare of CBA live material, inclusive of seed and broodstock. Storage and transport in confined areas or holding facilities, increased densities and reduced water quality, and poor handling, may, along with stress, increase the incidence of disease and infection and become an animal welfare concern. To maintain optimal health, it may be necessary to treat aquatic organisms to reduce stress and to prevent or eliminate disease or infections in the captured stock prior to arrival at the culture facilities. Reliable information is not often available on the most appropriate and effective treatment products and procedures in relation to specific stress, disease, infection or species needs.

Recent years have seen the introduction of many regulations regarding consumer protection and biosafety issues. These often restrict the use of chemicals such as anaesthetics (e.g. phenoxyethanol), antibiotics, antioxides and antifoam agents, all products useful in reducing fish stress and preventing or treating disease or infection but of concern to human health. Information is lacking on bio-accumulation of these pharmaceuticals in live material, safe levels for human consumption and the potential transmission to consumers through aquatic food products. Hence, procedures for their use should be precautionary.

Pharmaceutical treatments should be administered by authorized veterinarians. They should address stress, disease or infections during transport and handling from capture to captivity, should be appropriate and should not adversely affect human health and safety. In particular:

- Best available information should be obtained on the most appropriate and effective treatment in relation to the specific stress, disease, infection, or species situation.
- Pharmaceutical treatment plans should exclusively use authorized therapeutic agents and at the prescribed doses.
- Data on the incidence of disease and infection, and pharmaceutical treatments, should be collected, compiled and analysed.
- Treatments should be diagnosed and provided by suitably qualified persons.

3.2.1.3 Transfer of diseases, parasites and live material in relation to CBA

Capture-based aquaculture typically includes the transport of live material, inclusive of seed or broodstock, from the location of capture to the location of culture facilities. The movement of live material for CBA often involves: (i) significant quantities of live material; (ii) sustained periods of time (from days to weeks or months); and (iii) substantial distances (for example, when seed is traded internationally). The use of best practices in the treatment for disease, pathogens or parasites may not always be effective in preventing or controlling their spread and additional measures may be necessary.

The FAO technical guidelines on the Health Management for Responsible Movement of Live Aquatic Animals (FAO, 2007) provide comprehensive overall guidance on addressing issues in relation to both wild populations and aquaculture facilities. Codes of practice for the introduction of alien species by the International Council for the Exploration of the Sea (ICES) (ICES, 2005) and the European Inland Fisheries and Aquaculture Advisory Commission (EIFAAC) (Turner, 1988; see also Appendix 4) and the FAO technical guidelines on Genetic Resource Management (FAO, 2008b) provide guidelines and an example of how good practices, inclusive of transfer and quarantine, can be developed (Box 9).

The capture and movement of live material from one area to another, and its maintenance in systems at culture sites open to the ecosystem, present the potential for escapees and the introduction of disease, parasites or genetic material (eggs or seed in most cases, but include broodstock in some operations) into wild populations of the same species in other areas, or into wild populations of other related species. Cultured live material may be from

Box 9
Species transfers and introductions – codes of practice

Article 9.3 of the FAO Code of Conduct for Responsible Fisheries specifically calls for establishment of codes of practice at both national and international levels. It recommends that "States should, in order to minimize risks of disease transfer and other adverse effects on wild and cultured stocks, encourage adoption of appropriate practices in the genetic improvement of broodstock, the introduction of non-native species, and in the production, sale and transport of eggs, larvae or fry, broodstock or other live materials. States should facilitate the preparation and implementation of appropriate national codes of practice and procedures to this effect." Article 9.3.2 further recommends that "States should cooperate in the elaboration, adoption and implementation of international codes of practice and procedures for introductions and transfers of aquatic organisms." A code of practice has been established by the International Council for the Exploration of the Sea (ICES). It provides advice on how to reduce the risk of adverse effects from the intentional introduction of marine and brackishwater alien species. The general principles of the code also apply to freshwater ecosystems. This code has been adopted in principle by several FAO statutory bodies. The requirements start with the preparation of a proposal that will be reviewed by an independent body. This code would also apply to the movement of gametes, juveniles and adult organisms used in hatchery-based aquaculture and capture-based aquaculture.

the wild and/or from hatcheries. The introduction of disease, parasites and unwanted genetic material into culture facility stocks creates the potential for the spread of diseases, parasites or genetic material into hatcheries or other locations in the chain of custody, or at the location of the culture facility (Box 10).

During handling, transfer and transport of live material for CBA, the escape of live material prior to arrival in the culture facility and while in the facility, and the introduction of exotic eggs, seed, broodstock, disease or parasites into wild populations or into culture facilities, should be prevented. Furthermore:

- Best practices should be identified and implemented to prevent the escape of live material, in the form of eggs, seed or broodstock, and to avoid the transfer of pathogens or parasites to wild populations outside the capture area.

Box 10
Introduction of exotic species – eels

At the beginning of the 1980s, a parasite specific to the eel (*Anguillicola crassus*) was first detected in the European eel, *Anguilla anguilla*, following the introduction of the Japanese eel (*Anguilla japonica*) to the Mediterranean Sea and Germany. This roundworm inhabits the eel's swim bladder, the lining of which gradually loses its elasticity and flexibility, leading to lesions. This condition may impede highly contaminated silver eels from reaching the spawning areas or reduce the viability of the larvae produced.

- Best practices should be identified and implemented to avoid the transfer of pathogens or parasites from the wild-caught live material to the live material or broodstock already in culture facilities.
- Quarantine facilities and practices should be incorporated into the overall development of CBA when transfer and transport of live material is involved.
- Best practices should be identified and implemented to prevent the escape of live material and to avoid the introduction of genetic material to wild populations outside the capture or culture area.
- Best practices should be identified and implemented to avoid the introduction of genetic material to the live material already in culture facilities.

3.2.2 Culture and grow-out issues

3.2.2.1 Provision of appropriate aquaculture conditions for wild-caught live material

Capture-based aquaculture is based on the culture of live animals taken from wild populations, and, as a natural resource, it should be kept within biologically sustainable levels while also addressing equity issues if it forms part of a multisector fishery on the target species. To ensure that the most beneficial and least wasteful use is made of limited natural resources, good practices are needed during culture because the unnecessary or excessive mortality of CBA stock in the aquaculture phase will mean additional pressure on wild fisheries to replace the losses during aquaculture and results in lost opportunities and waste of live material. The production of less than optimal outputs, e.g. lower-quality product, or smaller-sized animals than are suitable for culture operations, may also result in higher pressure on wild stocks to provide additional live material. Good practices and guidelines are available for the development and operation of responsible aquaculture conditions, techniques, equipment and facilities appropriate to many of the species involved in CBA.

Aquaculture operations should ensure maximum survival and/or optimal production from wild-caught live material. Culturing operations with CBA live material should employ best practices, with appropriate conditions, techniques, equipment and facilities to ensure appropriate conditions for maximum survival and/or optimal production from wild-caught live material.

3.2.2.2 Health monitoring during acclimatization and weaning

Regular inspection and monitoring of the health of live material in CBA systems is important for practical, economic and animal welfare reasons, especially during acclimatization and weaning. Some wild-caught live material will not adapt well to captivity or will not readily be weaned onto the feed provided. These individuals will starve and/or not remain healthy. The situation can be improved by communication between CBA fishers and hatchery managers focusing on the earliest-stage larvae to be caught, or the stage or stages most suitable to successful acclimatization and weaning, if possible, and transferred to controlled weaning conditions that result in higher survival and improved economic viability, as well as better animal welfare.

The weaning of CBA live material and broodstock onto artificial feed is a critical aspect of aquaculture operations for many species of wild-caught live material. Capture-based aquaculture live material at the time of capture is accustomed to eating wild food and has not been weaned onto artificial feeds.

During weaning, mortality and injuries can occur and injured fish can be vectors for disease, in addition to there being animal welfare issues of injuries. To prevent the spread of disease and maintain the welfare of aquatic organisms, it is important to identify and remove injured live material stock as soon as possible and to address the causes of excessive injuries or mortalities.

During acclimatization and weaning of CBA live material, organisms should be regularly inspected for health, and injured, non-adapted and starving individuals should be removed as soon as possible and handled accordingly. Every effort must be made to address fish welfare considerations.

Mortalities that occur during or as a result of the acclimatization and/ or weaning process should be documented and the causes understood and addressed where possible. Excessive mortality should be a matter of concern. Mortality levels documented should be included when considering overall mortalities of CBA-sourced animals from capture and throughout the culture process and prior to sale as part of estimating the total take from the wild.

3.2.3 Broodstock issues

3.2.3.1 Fisheries providing wild live material for CBA: broodstock

The capture of broodstock for use in aquaculture is included in the definition of CBA because of the common practice of regularly catching wild broodstock for maintaining culture operations. In many places and for many species, it is likely that this practice will continue until HBA closes the life cycle on commercially important CBA species and the stock becomes domesticated. Even then, wild relatives in nature will be important as a back-up source of genetic material and as a valuable resource in its own right. Wild broodstock taken for CBA operations is a particularly important issue when there is significant adverse impact from such collection to support aquaculture, for example, when: (i) broodstock harvesting occurs on a repeated and regular basis and for each generation of production; or (ii) a stock is particularly vulnerable to even limited capture of broodstock, such as from unmanaged spawning aggregations; or (iii) the species is rare or threatened. Given that the aquaculture facility has the technical capacity to feed and spawn wild-caught fish and collect gametes or eggs, in general, using a limited number of broodstock to produce seed for aquaculture is relatively efficient, and broodstock take can be managed as part of a responsible fishery. However, care must be taken that sufficient genetic diversity exists within the cultured population to allow it to perform well under culture conditions. In some cases, broodstock may be used for culture operations and returned to the wild. In such cases, appropriate handling and release protocols are needed.

In practice, the use of broodstock to produce seed lends itself relatively easily to conventional management when there is good information on the fishery, capacity and enforcement. However, the case of threatened target species may need special attention. There are some aquaculture operations based on broodstock taken from the wild for which there is also a fishery (e.g. *Penaeus monodon*), and the broodstock fishery has a significant impact on the conventional fishery in these cases. Many aquaculture systems need to resort periodically to the use of wild broodstock to refresh the genetic diversity of captive broodstock or to replace broodstock. Such operations typically take fewer animals and should have a schedule of replacement defined.

Currently, only some fisheries for broodstock are likely to engender significant impacts and require specific management measures, e.g. because of the large numbers that are collected or because the species is of conservation concern. In such cases, it is particularly important that the CBA broodstock fishery is managed responsibly. However, as capture fisheries decline and the need for seafood increases, more pressure could be put on wild broodstock. Hence, fisheries for broodstock for CBA should be managed for sustainability within the overall context of all uses of the target species (i.e. both CBA and non-CBA uses). Special attention will be needed when the capture of target broodstock threatens a species or population, and in particular:

- The CBA broodstock fishery should be managed in relation to the biological and ecological sustainability of the overall stock and according to the Code. Special attention will be needed in the case of rare and threatened species or populations. While in general it is not expected that broodstock fisheries represent a significant component of fishing pressure on the species, this may not be the case for threatened species or future scenarios of scarce seafood in which case they would need to be carefully managed.
- Data on broodstock fisheries should be collected, compiled and analysed to identify those fisheries that are significant and require management. Particular attention is needed to collect detailed data on the capture of broodstock in the case of threatened or vulnerable species to ensure that the reproductive capacity of the wild target population is not being compromised.
- The wild population that is the source of broodstock should be monitored. If the fishing pressure on broodstock for CBA is high enough to affect recruitment to the fishery, then management is needed.
- Efforts should be made to maintain broodstock in culture operations in good condition if they are retained for repeat spawning in order to minimize frequent and large-scale replacements of broodstock from wild stocks. Broodstock may need to be replaced to meet genetic goals of diversity (particularly in relation to restocking or stock enhancement) and performance.
- For broodstock used in culture and then returned to the wild, suitable transfer and release protocols should be applied to keep mortality rates and risk of spreading disease to a minimum.
- Broodstock of threatened species or populations (e.g. according to IUCN or CITES listings or based on national-level assessments) will need careful management for sustainability.

3.2.4 Feed issues

3.2.4.1 Dependence on wild aquatic species as feed for CBA

Although HBA and CBA both use wild aquatic resources (either fresh or processed) as feed, this practice tends to be a prominent issue for many CBA species because many are carnivorous. There are a number of species that are still fed directly on fresh fish derived from capture fisheries, i.e. low-value fish or invertebrates, so-called "low-value/trash fish" (note that low value does not infer that the species are not without potential value for human food), or rely heavily on bait fish/forage fish. The wider ecosystem implications of removing a large and diverse range of species, either, indirectly as bycatch or directly as dedicated feed-fish fisheries, is often unknown. However, given that many bycatch species taken are food fish for other wild commercial species, and given the enormous volumes of wild fish taken for feed, the potential for ecosystem effects cannot be ignored (Box 11).

More broadly of concern in culture operations is the overall efficiency of the direct use of fresh, wild-caught aquatic organisms in terms of best use of biomass as aquaculture feed owing to their low food conversion ratios, as well as possible health risks arising from the practice. Today, there is a growing trend towards encouraging the use of formulated feeds that are less reliant on aquatic organisms from capture fisheries and/or fish meal and can be considerably cheaper in the long term.

For many CBA cultured species, commercially produced formulated feeds are not yet widely available, or may not currently be commercially attractive. Many compound feeds are still in the developmental stages with heavy reliance on wild-caught fish. Research is needed on the specific nutritional and palatability requirements of wild-caught aquatic organisms to develop cost-effective artificial feed that will ensure low mortality during weaning and the high meat quality required by the market. For some cultured species in developing countries, there remains insufficient research on compound feeds, and this area requires attention. Furthermore, research is also needed on the use of non-marine products (e.g. terrestrial animal by-products, meat meal, blood meal) as fish feed. However, there are often market and socio-economic pressures in relation to new feed types, e.g. the Japanese market is not so interested in tuna fed on artificial feeds because of taste and texture concerns (Box 12), while fish farmers in the Hong Kong Special Administrative Region prefer to use wild-caught fish rather than pelleted feed in their culture operations because it is cheaper and easier to obtain.

Guidelines on the use of wild fish as feed in aquaculture have been developed by FAO and are available as one of the supplements to technical guidelines on aquaculture development and provide comprehensive overall guidance on addressing this issue (FAO, 2011c).

Dependence of CBA species on wild-caught fish should be reduced as much as possible and eliminated whenever possible, and volumes and species composition of wild feed-fish should be documented. In particular:

- In fisheries where there is a harvest of wild aquatic organisms for use as aquaculture feed, assessment of sustainability should be undertaken and specific management regulations developed, employing best practices for management, handling and quality control of these feed-fish fishery products.

Box 11
Use of feed fish may promote overfishing of feed-fish fisheries – the switch to pellet feed is a challenge

Feed fish are extensively used for mariculture in Southeast Asia, mounting regional concerns over the general decline of fish stocks in the region. The Hong Kong Special Administrative Region has a small mariculture sector, mainly grouper species. Despite government attempts to convert the mariculture sector from dependence on wild feed-fish, aquaculturists have been reluctant to change (Chau and Sadovy, 2005). As a consequence, culture zones have pollution problems caused by excess feed use, and large volumes of feed-fish are involved. The Chau and Sadovy study recorded species composition, fish sizes and volumes involved in the fish-feed fishery of the Hong Kong Special Administrative Region and determined that at least 109 fish species from 38 families were involved, mainly Leiognathidae, Clupeidae, Apogonidae, Carangidae and Engraulidae. Mean lengths and weights of these fish were about 8 cm and 7 g and many of the fish had not attained the size of sexual maturation. The estimated volume of feed-fish used annually was about 9 700 tonnes in 2002 based on a mariculture production of 1 211 tonnes and a feed conversion ratio of 8:1. Such use of small fish for feed is not considered appropriate because: (i) its use exacerbates the pressure of overfishing in Hong Kong Special Administrative Region waters where there is little fisheries management; (ii) there are unknown effects on the marine ecosystem caused by the removal of large volumes of small pelagic fishes, as calculated from feed conversion ratios and fish production; and (iii) the mixed fish include many species that could be used for human consumption if allowed to grow larger.

Box 12
Challenges of developing formulated feed – the case of Mediterranean tuna

Bluefin tuna are fed mainly with a mixed diet composed principally of small pelagic species including sardine (*Sardinella aurita*), pilchard (*Sardina pilchardus*), herring (*Clupea harengus*), mackerel (*Scomber japonicus*), bogue (*Boops boops*) and squid (*Illex* sp.). Considering the high volumes of mixed fish needed, and heavy reliance on wild populations of feed-fish for tuna (2–10 percent of the bluefin tuna biomass farmed), there is an urgent need for research to develop artificial diets able to support a better feed conversion ratio and to ensure greater control over the quality of the fish produced. Scientific evidence indicates that fish weaned on a formulated diet that replicates normal nutritional intakes will perform considerably better than those fed on mixed fish-feed, eliminating health risks associated with the use of raw fish. High production costs and resistance from the Japanese market (owing to taste concerns) are problems yet to be overcome in adopting artificial feed. Challenges to progress in the development of feed for tuna include difficulties in working with large pelagic species (their high economic value making studies with live animals particularly expensive) and poor knowledge of the nutritional requirements of the species.

- Where CBA operations are dependent upon wild aquatic organisms for feed, research and development of artificial feed that reduces dependence of wild-caught feed should be promoted.
- The quantities, sizes and species of aquatic organisms serving as live feed for CBA species should be documented, as well as their origin, and efforts made to reduce bycatch that is not used in any way.

3.2.4.2 Transfer of disease, parasites or genetic material from CBA live feed

As with the capture and movement of live CBA material, inclusive of feed and broodstock, the capture, transport and use of wild feed-fish creates the potential for introduction of diseases and/or parasites into wild populations of the same species, or the populations of other species. The growing use of wild feed-fish for the CBA industry is increasing these risks. Treatments that reduce or eliminate pathogen load in feed are important to address this problem. The FAO technical guidelines for Health Management for Responsible Movement of Live Aquatic Animals provide comprehensive overall guidance on addressing these issues (FAO, 2007).

The introduction of diseases and parasites into wild populations from culture facilities and from the feed used should be prevented. In particular:

- Best practices should be identified and implemented to avoid the transfer of pathogens and/or parasites to wild populations outside the capture area of the CBA stock.
- Best practices should be identified and implemented to avoid the transfer of pathogens and/or parasites from the wild feed-fish to the live material already in culture facilities.
- In cases of risk of disease transfer from wild-caught feed, these organisms for feed in aquaculture should be treated to reduce this risk.

4. SOCIAL AND ECONOMIC CONSIDERATIONS

The social and economic benefits of CBA and HBA are often considerable, and HBA is not always preferable to CBA. In some situations, the collection and grow-out of wild juveniles and other live material for CBA provides considerable socio-economic opportunities to communities that HBA would be unlikely to provide. On the other hand, HBA can greatly improve culture operations by supplying a more constant and often more healthy seed, thereby standardizing production and reducing production risks. Seed shortages and quality issues can be a major constraint to aquaculture development, and breakthroughs in artificial breeding tend to lead to increased production (Box 13).

The gathering of seed from the wild at small scales and their sale to grow-out operators can generate significant employment and income for large sectors of the population otherwise excluded from the aquaculture industry and unlikely to be able to engage in hatchery production because of knowledge or financial constraints. Many hatcheries require considerable funding and advanced technology typically out of the reach of poorer sectors of society without government or other assistance. Seed fisheries and grow-out for CBA can support rural development and provide alternative or supplemental livelihoods. Operations are generally located in rural areas and can make considerable contributions to rural economies and social networks. This can result in significant economic multipliers within the economy through employment,

Box 13
Usage patterns of wild shrimp in Ecuador show advantages of HBA over CBA

During the 1970s and 1980s, the Ecuadorian industry relied almost entirely on wild shrimp post-larvae (PL). However, shortages of wild seed during the 1980s led to episodes of violence in the Ecuadorian estuaries ("post-larvae wars") (Csavas, 1994), while unpredictability in wild PL supply and disease outbreaks forced the industry to switch gradually towards hatchery PL. According to Sonnenholzner *et al.* (2002), records of the larvae sourced by 14 shrimp farms in Ecuador in the period 1995–2000 indicated a decrease in the number of ponds stocked with wild PL from 58 percent in 1995 to 7 percent in 2000. The switch from captured-based aquaculture (CBA) to hatchery-based aquaculture (HBA) was made possible because of technological and economic developments with resulting increases in shrimp production through more stable and healthy HBA supply of PL.

diversification of household livelihoods, small business development, purchase of goods and services, increases in income and food security, generation of foreign exchange and activities for women and children.

While there can be socio-economic advantages in relation to seed capture and supply for CBA, there can also be disadvantages associated with its practice that need to be considered in managing for long term sustainability of the fishery and clear advantages to promoting HBA (Box 14). Fishing methods used to take the seed can employ inappropriate technologies and skills, and users may undertake unsustainable practices to supply farmers with wild seed. Furthermore, the poor may be excluded from participating in or enjoying the benefits of wild seed collection and aquaculture production when large commercial interests become involved. Elites and/or politically powerful sectors may appropriate resources for themselves, or their activities

Box 14
An important local livelihood in the Philippines – the milkfish case

The milkfish, *Chanos chanos*, is one of the most important brackish-water, low-priced species cultured in Southeast Asia and an important food fish in the Philippines. However, supply of wild fry is declining owing to one or a combination of pollution, loss or degradation of coastal habitats and overexploitation of fry and/or adults, with both social and economic consequences. Although artificial reproduction is possible, HBA is not yet practised on a large scale, and thus most milkfish larvae used for grow-out in the Philippines are still taken from the wild. Moreover, associated with seed collection, bycatch and mortalities can be high and result in much wastage of target and non-target species. The Philippines currently imports 360 million fry in peak months. Demand for fry is increasing as a result of a shift from traditional or extensive culture systems to semi-intensive and intensive culture systems, and a move from prawn farming to milkfish farming following the collapse of the prawn farming industry. Because wild fry are still preferred by many milkfish producers, management is needed to reduce wastage, monitor trends over time, develop local participation and ensure equity and continuation of wild fry harvest. While hatcheries will be important for addressing the increasing demand for milkfish fry, this will mean competition for fry gatherers, requiring mechanisms to ensure the sustainable continuation of wild fry capture. Enforcement of rules and regulations for milkfish and fry smuggling is also needed.

may affect other fishing sectors, e.g. the taking of large amounts of seed of a species may negatively affect the fishery on adults of the same population. On the other hand, the development of HBA can generate additional jobs, ensure more stable seed supply and increase productivity. It might also help to address situations of conflict and violence stemming from competition for the target wild resource. Whether for CBA or HBA, there may be a need in many cases to develop and promote certification and best practices to ensure market access or to be sufficiently competitive.

Of critical importance for establishing productive, responsible and sustainably managed CBA culture operations are suitable management arrangements, enforcement, legislation, information, education and statistics, and consultation with stakeholders. Both the target fishery and the wider ecosystem context are relevant with the resulting practice (or practices) seeking a positive balance across all interested sectors and stakeholders.

4.1 Social considerations

A common key component in CBA, which typically focuses on early life history phases of wild-caught material (thereby, differing considerably from more conventional fisheries, which usually target older or adult stages), is the issue of equity of resource use and access by different fishing sectors. In some cases, this will also involve transboundary considerations, especially when species are highly migratory, as for pelagic species, or have different life history phases in different countries, as in many freshwater species. Even within a country, different fishing sectors may focus on different life history phases of the target species. In such cases, a major consideration should be the appropriate allocation of life history phases among different user groups in a way that takes into account their rights and needs within the overall context of the sustainability of the species. For example, removing too many adults could reduce the production of young recruits, thereby affecting the availability of fingerlings to seed fishers. Conversely, removing too many fingerlings could reduce the number of adults available to another fishing sector. In some cases, too much catch in the commercial sector might reduce animals available to the recreational sector. Moreover, if private interests monopolize key fishing grounds, use harmful fishing practices or generate bycatch wastefully, other resource users may be unjustly excluded or negatively affected (Box 15). Additional examples are provided in Boxes A3.4 and A3.5 in Appendix 3 (cases studies on tuna and carp, respectively).

Especially for areas with depressed and marginal economies and where employment opportunities are limited, the alternative fishery-related activities generated by CBA are often welcome alternatives for the existing fishing workforce. Employment opportunities are also made available in aquaculture production and marketing, and there are possibilities for fishers to become active partners in aquaculture activities either as suppliers of seed or as grow-out farmers.

Because target resources may not be able to accommodate all the social needs and pressures on them, consideration of the best and most practical overall use of the resource is required, with considerations for both equity and sustainability following the principles and recommendation in the Code

Box 15
Socio-economic impacts in the shrimp seed fishery in Bangladesh – lack of alternative options

In Bangladesh, the demand for shrimp post-larvae (PL) for capture-based aquaculture (CBA) increased in the mid-1980s with the rapid expansion of the shrimp industry. The resulting growth in fishing pressure on the fry fishery is thought to be contributing to declines in abundance and distribution of mother shrimp, thereby causing serious damage to the productivity of the coastal marine fishery sector that targets market-sized shrimp. In addition to fishing pressure, huge numbers of eggs, larvae and juveniles of non-target fish and shrimp harvested during shrimp fry collection are taken as bycatch. Overfishing of these fisheries has occurred to such an extent that fishing in the artisanal sector is no longer remunerative, with the fry fishery removing an estimated 90 percent of the *Penaeus monodon* fry stock. There are also concerns about the adverse impact of wild shrimp PL collection on wetland biodiversity. Despite a ban on wild shrimp PL collection since 2000, the practice continues to be a lucrative livelihood option for thousands of poor households living in coastal areas. The preference for wild shrimp PL for farming compared with hatchery-produced ones makes wild PL collection a profitable occupation. Most PL collectors come seasonally from other areas for the harvest, and most are from destitute non-fisher households. The profitability of wild PL harvest and lack of livelihood alternatives for shrimp PL fishers have made it difficult to remove and relocate PL collectors. Alternative livelihood options, such as small trades, fish trading and handicrafts, require training and credit support. Government projects to move collectors to alternative income-generating options have been unsuccessful as there are huge numbers of people engaged in this practice in remote areas of the coastline where it is difficult to reach and motivate people and to monitor activities.

guidelines. In particular, equity considerations call for the interests of all the stakeholders being properly represented, the livelihoods of local people being protected, and the well-being of disadvantaged groups being seen as a priority (Box 16).

In order to ensure the equitable distribution of costs and benefits of developing CBA operations, it is important that stakeholders depending on the resource being exploited are identified; issues related to food security, gender, cultural practices, existing tenure and user rights systems are carefully considered; and possible user conflicts are identified and addressed.

4.2 Economic considerations

The establishment, management and monitoring of sustainable CBA activities require sufficient and ongoing funding and, in the early stages, a research and development phase may be needed. It will also be necessary to provide

Box 16
Questions over best use of mullet resources exposed to multiple fishing sectors in Egypt and beyond

Fishery regulations of wild mullet resources in Egypt for their best possible overall economic and social benefits is a major challenge given their exposure to multiple fishing sectors, i.e. for roe, seed and adults. One concern has been declining seed availability. However, despite these declines, the continued availability of cheap wild seed has meant that private investors have not yet been sufficiently attracted by government incentives to invest in hatcheries. A different fishery is the harvesting of mature mullets during their spawning migrations from coastal lagoons, lakes or rivers to the sea to extract the ovaries. The fishing of ripe mullet for roe production practised for centuries in the Mediterranean region, has extended to other regions in the last few decades, especially in Asia and the United States of America. Mullets are known to be highly fecund and one kilogram of fresh ovaries is equivalent to the loss of 10–15 million eggs. The losses to recruitment of wild mullet evidently associated with the roe fishery are estimated to be at least 6–8 times that associated with seed collection for capture-based aquaculture (CBA) of mullet. A ban on collection of wild seed may stimulate development of hatchery production of mullet seed, but its effects on the growing activity of aquaculture must be considered.

training and establish sufficient capacity in terms of human resources. Economic analyses may also be required to determine the market situation or requirements (e.g. through certification of good practices), the economic impacts of a CBA operation on other wild fishery sectors (with consideration of possible support or financial compensation for disadvantaged fishery sectors), for ongoing monitoring of catches and fishing activity, and to address possible chain-of-custody considerations.

In the development of CBA activities, socio-economic considerations, in addition to sustainability and equity, are necessary. Funding should be identified and committed to ensure long-term sustainability for the various developmental phases, from research, to fishery, to trade and economic analyses, and to training, monitoring and enforcement. Furthermore:

- A cost-benefit analysis should indicate that the benefits to society outweigh the costs of a CBA operation.
- The CBA operation should adequately address sustainability and social equity issues, including environmental externalities, and be financially viable at different scales and fishing levels of operation.
- An economic sensitivity analysis in relation to other uses of the stock, and other uses and/or stakeholders should indicate an acceptable level of balance between the CBA use of the stocks and other uses, such as fisheries on market-sized adults, including impacts on existing resource users.
- Long-term funding will be necessary to establish management plans and enforcement needs, as well as for monitoring and data collection to enable adaptive and effective changes in the management plans.

5. GOVERNANCE CONSIDERATIONS

Where a decision is made that an existing or proposed CBA activity is promising or suitable, the following section provides guidance on ensuring that the wild capture fisheries are responsible, with a particular focus on aspects that are unique to CBA. It covers management arrangements, effectiveness and compliance, legal and institutional frameworks, information and statistics, as well as communication and/or consultation with stakeholders. It also addresses the operation of both fisheries and aquaculture in the special case of threatened species.

5.1 Holistic management approaches linking fisheries to aquaculture

Capture-based aquaculture depends on capture fisheries for seed and/or broodstock, as well as for feed in some cases. Therefore, it must: (i) be operated or developed according to responsible and sustainable fisheries principles and practices; (ii) consider equity issues; (iii) respect conservation and management measures; and (iv) address animal welfare, i.e. accommodate an EAF (Box 17). Hence, appropriate legislation and governance are essential.

Box 17
Cod legislation for licensing of equipment and welfare in Norway

Although capture-based aquaculture (CBA) of cod in Norway dates back to at least the 1880s, specific legislation that recognized the hybrid nature of CBA between fisheries and aquaculture did not evolve until 2006. Previously, operations were organized separately, with harvest governed through the fisheries act and fish farming through aquaculture acts. Only within the last decade have authorities considered the need to bridge regulations across the two types of operation to allow for better governance and for economically sustainable development of the CBA sector. Resource control and animal welfare issues have been central to the development of new legislation. Catch of cod for CBA is restricted by minimum size and quotas to avoid overfishing and must be fully documented. Fishing vessels must report in advance when they start fishing for CBA and when they deliver the catch, which is inspected. Animal welfare considerations specify that fishing vessels be appropriately equipped for transferring fish from the fishing gear to the boat, for sorting and moving fish into the holding tanks, and for ensuring welfare in terms of fish density, water flow and transport times. Fish must be checked by a veterinary expert, with injured animals being killed immediately.

Capture-based aquaculture is likely to continue in the long term for many species even when HBA is, or becomes, possible and where there may be considerable social advantages to continuing CBA (e.g. Box 13). Moreover, wild broodstock may periodically need to be captured to maintain genetic diversity or to replace dead broodstock. It is clear that appropriate governance of CBA fisheries should be given a high priority (FAO, 1995).

5.2 Development of management plans for CBA-related fisheries

Management intervention should be as effective, practical and cost-efficient as possible. It must also be developed around clear objectives for the fishery, appropriate reference points (or management targets) identified (FAO, 1995), and involve documentation and adaptive management (Cochrane and Garcia, 2009). These are the fundamental elements of a management plan.

For any CBA-related fishery, a management plan is required to identify key ecological, social and economic issues relevant to the sustainability of that fishery. The plan will also include suitable control measures to address, among others, animal welfare, minimizing mortality during capture, transfer and grow-out. Ecosystem and transboundary considerations will also be included, using the best available knowledge and precaution within an adaptive management framework.

The Code affirms that "States should apply the precautionary approach widely to conservation, management and exploitation of living aquatic resources in order to protect them and preserve the aquatic environment. The absence of adequate scientific information should not be used as a reason for postponing or failing to take conservation and management measures" (FAO, 1995).

Where CBA-related fisheries target highly migratory species and/or transboundary fisheries and the live material or broodstock originates from outside national waters, they may not be subjected to a national fishery management plan. This creates special management challenges. It is the responsibility of the country within which the aquaculture operation will be undertaken to report on the usage of the stock from international waters in the case of stock obtained from the high seas, e.g. highly migratory or shared stocks.

The development and conduct of CBA operations should be undertaken in accordance with the principles listed in Section 2.4.

5.2.1 Developing a management plan

As a first step, adequate information for management decision-making should be collected by the appropriate stakeholder(s), inclusive of biological research on natural resources, documentation of catches, social and economic aspects of the fishery, etc., as background for the planning and implementation processes. Objectives should be agreed both in relation to ecological sustainability, but also in relation to the social and economic achievements that the activity is meant to achieve.

Through a comprehensive participatory process, key issues related to ecological, socio-economic and governance aspects of the activity should be identified and prioritized keeping in mind the agreed overall objectives. Based on the above, an action plan is prepared and agreed by stakeholders, including appropriate management measures and needs for data and information. Transboundary issues or considerations related to multiple uses and users of the resource should also be clearly identified, and related implications for science and management evaluated.

Management plans can be developed following the process encouraged for the application of the EAF and EAA. To achieve the objectives of a management plan successfully and implement the actions foreseen in the plan, two elements are fundamental in the process: (i) to collect and use the best available information; and (ii) to have broad stakeholder participation. The process and steps for the development of a management plan are described in Figure 3.

Figure 3
Planning process and implementation of an adaptive CBA management plan

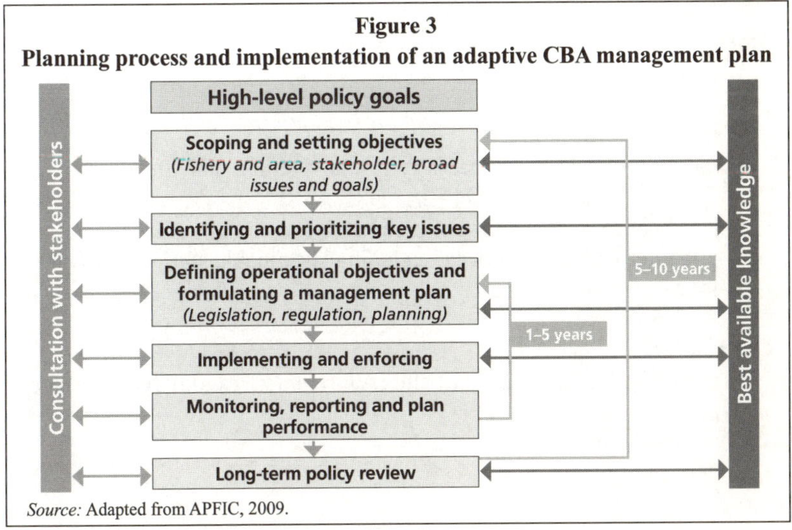

Source: Adapted from APFIC, 2009.

5.2.2 *Management tools for capture-based aquaculture*

A wide range of tools and measures can be applied to address the various challenges of managing the different components of capture-based aquaculture (refer to FAO technical guidelines on fishery management and on aquaculture). Examples include:

Input controls

- gear fees;
- licences for collection of seed or broodstock;
- restrictions or bans on certain fishing gear or modes of fishing.

Time/area closures

- seasonal ban in taking seed or broodstock;
- zoning of areas of biological importance to wild caught seed or broodstock;
- closed areas (seasonally or permanently);
- protection of key seed settlement or nursery habitats.

Output controls

- restrictions on the harvest of spawning adults;
- harvest locality limitations;
- minimum or maximum sizes for harvested species;
- catch allocations between fishing sectors (i.e. on seed, broodstock, conventional fishery on market-sized fresh fish).

Market-related measures

- export controls;
- quotas;
- traceability of product;
- industry codes or standards (e.g. the International Standard for the Trade in Live Reef Food Fish; see Box 18);
- certification systems for culture phase;
- good mariculture practices, including sustainable sourcing and use of feed.

Other measures

- improved harvest, transport and culture practices to reduce wastage;
- pollution controls;
- use of chemicals;
- control of disease;

- licence for transport, hatchery, transfer/transport of seed of broodstock;
- alternative livelihoods by those affected by a ban;
- permanent ban on seed or broodstock capture in the case of unsustainable practices, threatened species or allocations concerns.

The management of CBA fisheries is best implemented with the support of effective monitoring and documentation of catches and regular reporting on key aspects of the fishery. Ideally, the catches themselves should be recorded or logged rather than just the numbers of seed entering culture facilities. This is because, in the case of live catches, there may be significant mortality between the time of capture and the time of landing or entering culturing facilities. Important sources of mortality should be known because not only are they a component of fishing mortality, but, by recognizing them, mitigation measures may be developed through better fishing, handling or transfer practices. Moreover, animal welfare issues will need attention on some CBA-related fisheries.

Box 18

Adopting voluntary standards for good aquaculture practices relevant to CBA – an example from the trade in live reef food fish

Understanding and addressing the impacts and means of removals associated with target and non-target CBA species on biodiversity and the marine ecosystem, in addition to the various implications of international trade in live seed, is extremely important to ensure sustainable and responsible fisheries and associated trade, shipment and marketing practices. A voluntary international standard for the trade in live reef food fish was developed following extensive stakeholder consultations. Among other objectives, the standard aims to ensure that the "seed" taken to supply grow-out operations is produced on a sustainable basis while minimizing negative impacts on biodiversity and ecosystem caused by overfishing, damaging or wasteful capture methods. The standard also includes recommendations to address the shipment and transfer of seed, including practices to avoid the introduction of exotics through unwanted escapes or releases and the spread of disease. Although the standard is voluntary, it represents a comprehensive code of practice of direct relevance to CBA and could be considerably strengthened and supported by relevant local and national laws and by international agreements (see www.livefoodfishtrade.org).

Best practices for the management of both the aquaculture and fishery components of CBA including legal, regulatory, information and statistics aspects are being developed and implemented, or are in place, at different locations and for different species; they may provide useful guidelines in the development of new management measures (FAO, 1995; Cochrane and Garcia, 2009).

5.3 Legal and institutional framework

Aquaculture operations in general face multiple challenges, particularly where significant CBA activities are involved. This is because both HBA and CBA considerations need to be addressed. These range from the need to consider sustainable practices in relation to fisheries for feed, seed, broodstock, equity of resource use and the condition of the culture environment (e.g. water quality, pollution, disease), to considerations on the use of chemicals, traceability and animal welfare.

Aquaculture activities that involve the removal of live material from the wild may require novel and specific management intervention and legislation to be established, or reviewed, and research conducted. Funding may need to be generated or identified to ensure adequate management, including enforcement capability, and research capacity. A major challenge to address is the prevalence of illegal, unreported and unmonitored trade in many CBA fisheries (Box 19; see also Boxes A3.6 and A3.7 for additional information on eel and mullet in Appendix 3).

To ensure the long-term biological sustainability of CBA operations and the best possible social and economic benefits, appropriate information and management capacity are essential and enforcement must be effective. All CBA operations have to be legal, reported and regulated. Adequate capacity and funding are needed to address legislation, regulatory and institutional structures for managing CBA operations, and for monitoring, control and surveillance.

5.4 Trade considerations

On occasion, CBA will be conducted using species that are considered to be threatened. This might be in the context of population recovery and part of a specific conservation programme, or CBA may be associated with sustainable international trade under CITES. For threatened species under CBA that are listed on CITES Appendix I or II, there is a need to be able to demonstrate that

Box 19
Addressing illegal bluefin tuna seed fisheries

Illegal, unreported and unregulated (IUU) fishing is a major challenge to the sustainable management of Atlantic bluefin tuna stocks, and for tuna fisheries in relation to tuna capture-based aquaculture (CBA). Major problems include tuna transshipments at sea, the absence of data on fish weight at capture, uncertainty about information from fish fattening operations and in relation to reported national statistics, and data on international trade. Such shortcomings lead to underreporting and compromise stock assessments leading to a likely surpass of the total allowable catch (TAC) levels. Illegal, unreported and unregulated fishing activities in relation to CBA tuna activities include companies with their own, or affiliated, fishing fleets or tugboats; vessels flagged to different countries providing fish to the same farm in a country different to that of the fishing vessels; and reflagging of fishing vessels. All tuna-ranching countries in the Mediterranean import bluefin tuna caught by other countries to stock their farms, and most countries catching bluefin tuna are also engaged in their transport from the fishing ground to the farming location. This made it difficult to carry out controls at the beginning of tuna CBA activities in the 1990s. There has since been a concerted and productive international political effort to identify and address the problem of IUU fishing for bluefin tuna by regulating the fisheries and CBA tuna activities (observers, declarations of catches, transshipment at sea and to the farming cages, etc.).

any wild caught fish exported are done so based on a sustainable management plan and that any HBA component is clearly distinguished (e.g. by tagging) from CBA production. For commercial species listed on CITES Appendix II, the CITES calls for exports of the species to be conducted on a sustainable basis. Challenges in attaining sustainable use in such cases are twofold. The first is that threatened species are usually not common and may be little understood in terms of their fishery. The second challenge is that conservation issues in relation to commercially exploited marine species of fishes and invertebrates are relatively new concerns and are not typically handled by government fishery departments, and hence governance and enforcement can pose institutional challenges. Examples of threatened marine fishes for which both CBA and HBA are, or are likely to become, production means into the long-term future include seahorses, bluefin tuna and the humphead wrasse (Box 20).

Box 20
CBA implications for a threatened marine fish regulated in international tarde

Some species taken for capture-based aquaculture (CBA) are listed as threatened by the International Union for Conservation of Nature (IUCN) on its Red List of Flora and Fauna and/or included on one of the appendixes of the Convention on International Trade in Endangered Species of Wild Fauna and Flora (CITES). One example is the humphead wrasse, or Napoleon fish, *Cheilinus undulatus*, listed as endangered on the IUCN Red List in 2004 and also on the CITES Appendix II in the same year. This relatively uncommon species is widely traded in Southeast Asia at a preferred market size of 0.5–1 kg, at or below the size of sexual maturation. The species cannot be produced using hatchery-based aquaculture (HBA) at commercial levels and is mainly produced by wild capture of market-sized fish and CBA. If HBA develops, there will need to be a means of distinguishing wild-caught fish from hatchery-produced fish if both are included as part of the export quota. To produce a management plan, collaboration was established between FAO, the Indonesian Institute of Sciences and the IUCN Groupers and Wrasses Specialist Group. A fishery model was developed following collection of data from wild populations, grow-out operations and trade data. Indonesia and Malaysia applied the model to assess their humphead wrasse populations. Major challenges under discussion are with illegal, unreported and unregulated trade, which is undermining sustainable trade, and with institutional capacity for managing threatened marine fishes. This example demonstrates the value of collaborative studies and of following through with management initiatives to ensure they are effectively implemented (Sadovy *et al.*, 2007). It also illustrates the importance of considering CBA in determining sustainable export levels in compliance with the CITES.

For species used in CBA or HBA that are listed on the CITES Appendix II, careful monitoring of wild capture and of culture operations, and clear identification of HBA-sourced organisms, are needed as part of sustainable management (known as non-detriment findings under the CITES) in relation to export of the species.

5.5 Information, statistics and communication

For the purposes of understanding the relationship between CBA and wild natural resources, as well as for appropriately documenting aquaculture production, care is needed in monitoring the catch and effort associated with seed acquisition, and the relative production from CBA versus HBA culture operations. Countries reporting on fishery production need to be able to disaggregate data between wild fisheries capture for CBA on the one hand, and between aquaculture production derived from CBA and HBA on the other. The identification of the source of production is extremely important for clarifying the relative contribution to the overall food production by each type of operation. Moreover, cross-boundary sharing of information for aquaculture planning, development and reporting can assist with management planning for transboundary or migratory species. It is also relevant for other important issues such as helping to minimize the spread of disease in transferring animals across national boundaries, or the introduction of exotic species (Article 9.2 of the Code).

Examples of data requirements include the level of effort employed in the fishery (e.g. number of fishing days), types of fishing gear and, number and size classes of live material introduced into grow-out and broodstock holding facilities, losses through mortality at different stages from capture during transfer and grow-out. As for fisheries or aquaculture operations, it is important to monitor not only production, but also social and economic aspects of the operations to support an EAA and EAF. It is recommended that:

- A process should be in place that ensures collection of information and data on wild seed/broodstock removal for CBA practices and for sound management of the relevant wild resources.

5.5.1 Traceability and documentation of CBA live material from capture to culture

Tracking the movement and status of live material, inclusive of seed and broodstock, as it moves from the point of capture through to culturing facilities is an important component of responsible CBA and necessary for the documentation of species volumes and sizes involved for use in stock assessment and adaptive management. It is important to be able to distinguish and document the portion of stock removed from the wild for CBA fisheries, including mortality levels at different stages of handling and transport of captured live material, as well as the contribution of CBA live material to total aquaculture production by species and at national and global levels. The amount of production that occurs during the culture phase of CBA can be documented as aquaculture production. However, careful consideration is needed regarding how to document the capture fishery component of CBA-related fisheries. Two general classes of capture fishery can be considered: (i) when the seed stocked are very small and relatively few in number, then documentation may not be necessary unless the species is threatened, in which case all capture should be recorded; and (ii) documentation of wild capture (by size and volume) should be included in national landing figures in the case of large catches of animals for CBA, as the information may be needed for management plans or quota allocations. Food safety and sustainability issues also increasingly require that that there be credible traceability information on the origin of aquaculture products and the practices employed along the chain of custody (Box 21; see also A3.8 for a case study on tuna in Appendix 3).

Addressing these considerations requires traceability of CBA live material from the culture facility back to the fishery, as well as the documentation of the numbers and sizes of animals involved at capture and during culture if significant numbers are removed or if the species is threatened. Information on the movement of the live material can help with the preparation of statistics for management and regulatory purposes. Where data on the fishery for the live material are unavailable or unreliable, and a significant portion of the CBA live material is exported, export statistics can give a basis for estimation of catch. In this case, care may be needed to distinguish live material being exported that derives from CBA from that derived from HBA. Therefore, it is recommended that:

- Capture-based aquaculture operations should include documentation systems for traceability by recording the movement and mortality (or loss) post-fishing of live material at appropriate steps in the chain of custody.
- At a minimum, the collection and/or export of CBA live material should be documented.
- Traceability documentation systems for CBA, as distinct from HBA of the same species if applicable, should be standardized and harmonized as best as possible with capture fisheries, aquaculture and export/ import statistics and information systems.

Countries should, therefore, collect separate data on aquaculture production in terms of catches taken from the wild during CBA fisheries and grow-out in captivity; this is to clarify the relative contribution to production of each part of the operation (i.e. as fisheries or aquaculture production). Special attention is needed in the case of threatened species.

5.5.2 Communication and consultation with stakeholders

Communication and consensus building are essential components of successful development and implementation of fishery management. Consultation across

Box 21
Problem of grouper juvenile export and re-export and the need for traceability

Grouper seed of a wide range of sizes and ages, from shortly post-settlement through to subadult phases, is captured throughout much of Southeast Asia. Hundreds of millions of seed probably entered international export trade annually in the 1990s. Given the large number of countries involved and the many collecting sites, fishery or trade monitoring at the species level is a major challenge and is extremely limited. Therefore, not only are catch rates, fishing effort and fish sizes in most capture-based aquaculture (CBA) fisheries unknown, losses from mortality at all stages are not documented. Moreover, the ability to control for disease in imported fish and to trace back to the actual origin of traded fish appears to be non-existent. In addition, the consolidation of seed from many locations by traders masks declines in populations at the local level and can result in serial overfishing, which is likely to go undetected at both national and international levels for a long time.

different stakeholders (e.g. capture fishery operators and, if relevant, operators in other fishing sectors targeting the same stock, broodstock collectors, aquaculture operators and fishery managers) can facilitate the development of management approaches in a way that minimizes possible conflicts and fosters acceptance and compromise (see Figure 3). Estimation and recording of the biomass of exploited CBA species can help ensure that appropriate procedures and regulations are established, enabling effective management and monitoring of CBA resources by the relevant stakeholders. It would also enable adequate analysis on the effectiveness of adopted management measures and allow for adaptive management.

5.5.3 Increasing communication and public awareness

Educational campaigns and initiatives are important for building consensus and acceptance of management and behaviour change. They can also help to improve compliance and to make different public sectors more aware of natural resources issues (Box 22) (FAO, 2009).

A better understanding by the general public of the fisheries and aquaculture issues associated with CBA and in relation to sustainable use and the need for management is important for gaining public support for policies.

Box 22

Communication with the public – European eel information kit

Given concerns over the status of European eel stocks and recognizing the diverse nature of European catchments where the species occurs, management plans for the species are being conducted across its distribution area. Eel populations are not only affected by fishing; other problems include the use of water for energy and exploitation of wetlands for urban and agricultural needs. This range of problems means that improvement measures for habitats in one catchment area alone cannot lead to a restoration of the resource on a European-wide spatial scale, given the diverse structure of the population. To assist in the effective implementation of management plans and actions for this threatened species, educational materials in multiplc languages have been prepared on the biology of the species, on the major anthropogenic impacts affecting the future of the stock, and on the importance of protecting the environment. Furthermore, such communication material explains in easy terms the basis of a restoration plan.

6. FUTURE DEVELOPMENTS

Aquaculture continues to grow rapidly and provides an increasing proportion of fishery products for human consumption. Its relative importance in meeting aquatic protein needs will grow if global production from wild harvest remains stable at the current output level or decline. Aquaculture can take several forms depending on the type of culture system and the degree of dependence on wild fisheries resources. In the case of CBA, as there is a heavy reliance on the catch and culture of wild resources, both fishery management and aquaculture practices need to be considered. The Code provides guidance and principles relevant to good fishery and aquaculture practices. The EAF and the EAA provide frameworks for the comprehensive and practical implementation of the Code. They identify factors contributing to the resilience of both social and biophysical systems associated with the culture phase of CBA. Together, these sets of guidelines provide means to plan and manage aquaculture development in a way that integrates it with different fishery sectors and with HBA while aiming to ensure sustainable use of wild resources and consider overall social and economic benefits.

Two major challenges for CBA are the management of associated fisheries and attaining the appropriate balance between HBA and CBA once HBA has been achieved and where CBA-associated harvest supports significant livelihoods. The integration and coordination of different government sectors, across a range of fishery and aquaculture interests will also require the establishment of institutional mechanisms for effective coordination, as it will be necessary to consider human health, traceability and transboundary issues. Given the growing awareness of the need for an ecosystem-based approach to management, the development of high consumer expectations and the call for sound aquaculture practices, it is in the best interests of businesses, operators and governments to begin to factor in international standards and take account of the potential power of consumer preferences.

The CBA guidelines provide a general framework for the development and implementation of a biologically and environmentally sustainable culture sector that takes full account of ecosystem and social limits, as well as the interests of resource users and stakeholders. In moving towards better practices, there is a need to couple and integrate science, policy and management at the government level by developing specific policies, strategies and funding mechanisms. These guidelines should be considered as a work in progress, with potential for expansion, alteration and development in the future.

REFERENCES

APFIC. 2009. *APFIC/FAO Regional consultative workshop "Practical implementation of the ecosystem approach to fisheries and aquaculture"*, 18–22 May 2009, Colombo, Sri Lanka. FAO Regional Office for Asia and the Pacific, Bangkok, Thailand. RAP Publication 2009/10. 96 pp.

Beveridge, M.C.M., Ross, L.G. & Kelly, L.A. 1994. Aquaculture and biodiversity. *Ambio* 23: 497–502.

Chau, G.T.H & Sadovy, Y. 2005. The use of mixed fish feed in Hong Kong's mariculture industry. *World Aquaculture* 36(4): 6–13.

Cochrane, K.L. & Garcia, S.M., eds. 2009. *A fishery manager's guidebook*. Second Edition. Rome, FAO, and Wiley-Blackwell. 536 pp.

Csavas, I. 1994. Important factors in the success of shrimp farming. *World Aquaculture*, 25(1): 34–56.

FAO. 1995. *Code of conduct for responsible fisheries*. Rome. 41 pp. (also available at www.fao.org/docrep/005/v9878e/v9878e00.htm).

FAO. 1996. *Precautionary approach to capture fisheries and species introduction*. FAO Technical Guidelines for Responsible Fisheries. No. 2. Rome. 54 pp. (also available at ftp://ftp.fao.org/docrep/fao/003/W3592e/W3592e00.pdf).

FAO. 1997. *Fisheries management. 4*. FAO Technical Guidelines for Responsible Fisheries. No. 4. Rome. 82 pp. (also available at www.fao.org/docrep/003/w4230e/w4230e00.htm).

FAO. 2003. *Fisheries management. 2. The ecosystem approach to fisheries*. FAO Technical Guidelines for Responsible Fisheries. No. 4, Suppl. 2. Rome. 112 pp. (also available at www.fao.org/docrep/005/y4470e/y4470e00.htm).

FAO. 2007. *Aquaculture development. 2. Health management for responsible movement of live aquatic animals*. FAO Technical Guidelines for Responsible Fisheries. No. 5, Suppl. 2. Rome. 31 pp. (also available at www.fao.org/docrep/010/a1108e/a1108e00.htm).

FAO. 2008a. *Inland fisheries. 1. Rehabilitation of inland waters for fisheries*. FAO Technical Guidelines for Responsible Fisheries. No. 6, Suppl. 1. Rome. 122 pp. (also available at www.fao.org/docrep/011/i0182e/i0182e00.htm).

FAO. 2008b. *Aquaculture development. 3. Genetic resource management*. FAO Technical Guidelines for Responsible Fisheries. No. 5, Suppl. 3. Rome. 125 pp. (also available at www.fao.org/docrep/011/i0283e/i0283e00.htm).

FAO. 2009. *Information and knowledge sharing*. FAO Technical Guidelines for Responsible Fisheries. No. 12. Rome. 97 pp. (also available at ftp://ftp.fao.org/docrep/fao/011/i0587e/i0587e00.pdf).

FAO. 2010. *Aquaculture development. 5. Ecosystem approach to aquaculture.* FAO Technical Guidelines for Responsible Fisheries. No. 5, Suppl. 4. Rome. 53 pp. (also available at www.fao.org/docrep/013/i1750e/i1750e.pdf).

FAO. 2011a. *International guidelines on bycatch management and reduction of discards. Directives internationales sur la gestion des prises accessoires et la réduction des rejets en mer. Directrices internacionales para la ordenación de las capturas incidentales y la reducción de los descartes.* Rome/Roma, FAO. 2011. In press.

FAO. 2011b. *Report of the Technical Consultation to Develop International Guidelines on Bycatch Management and Reduction of Discards.* Rome, 6–10 December 2010. FAO Fisheries and Aquaculture Report. No. 957. Rome. 32 pp. (also available at www.fao.org/cofi/24783-010c9c0c7cae3b0bb7f6b70baec897306.pdf).

FAO. 2011c. *Aquaculture development. 5. Use of wild fish as feed in aquaculture.* FAO Technical Guidelines for Responsible Fisheries. No. 5, Suppl. 5. Rome. 79 pp. (also available at www.fao.org/docrep/014/i1917e/i1917e00.pdf).

International Council for the Exploration of the Sea (ICES). 2005. *ICES code of practice on the introductions and transfers of marine organisms 2005.* Copenhagen. 30 pp. (also available at www.ices.dk/reports/general/2004/ICESCOP2004.pdf).

Mous, P.J., Sadovy, Y., Halim, A. & Pet, J.S. 2006. Capture for culture: artificial shelters for grouper collection in SE Asia. *Fish and Fisheries*, 7: 58–72.

Ottolenghi, F., Silvestri, C., Giordano, P., Lovatelli, A. & New, M.B. 2004. *Capture-based aquaculture: the fattening of eels, groupers, tunas and yellowtails.* Rome, FAO. 308 pp. (also available at ftp://ftp.fao.org/docrep/fao/007/y5258e/y5258e00.pdf).

Rosser, A.R. & Haywood, M.J., compilers. 2002. *Guidance for CITES Scientific Authorities: Checklist to assist in making non-detriment findings for Appendix II exports.* Gland, Switzerland, and Cambridge, UK, International Union for Conservation of Nature. xi + 146 pp.

Sadovy de Mitcheson, Y. 2009. Biology and ecology considerations for the fishery manager. *In* K.L. Cochrane & S.M. Garcia, eds. *A fishery manager's guidebook*, pp. 21–51. Rome, FAO, & Wiley-Blackwell. 526 pp.

Sadovy, Y., Punt, A.E., Cheung, W., Vasconcellos, M. & Suharti, S. 2007. *Stock assessment approach for the Napoleon fish,* Cheilinus undulatus, *in Indonesia: a tool for quota setting for data-poor fisheries under CITES Appendix II Non-Detriment Finding requirements.* FAO Fisheries Circular No. 1023. Rome, FAO. 71 pp.

Sonnenholzner, S., Massaut, L., Saldias, C., Calderón, J. & Boyd, C. 2002. *Case studies of ecuadorian shrimp farming.* Report prepared under the World Bank, NACA, WWF and FAO Consortium Program on Shrimp Farming and the Environment. Work in Progress for Public Discussion. Published by the Consortium. 55 pp.

Turner, G.E., ed. 1988. *Codes of practice and manual of procedures for consideration of introductions and transfer of marine and freshwater organisms (1989)*. EIFAC Occasional Paper No. 23. Rome. FAO. 46 pp.

United Nations. 1992. *Report of the United Nations Conference on the Human Environment.* Stockholm, 5–16 June 1972. Rio Declaration on Environment and Development. The United Nations Conference on Environment and Development. New York. USA. (also available at www.unep.org/Documents.Multilingual/Default.asp?documentid=78&articleid=1163).

Appendix 1

GLOSSARY OF DEFINITIONS

Already defined terms

Aquaculture is the farming of aquatic organisms: fish, molluscs, crustaceans, aquatic plants, crocodiles, alligators, turtles and amphibians. Farming implies some form of intervention in the rearing process to enhance production, such as regular stocking, feeding, and protection from predators. Farming also implies individual or corporate ownership of the stock being cultivated. For statistical purposes, aquatic organisms harvested by an individual or corporate body that has owned them throughout their rearing period contribute to aquaculture. Because the original definition of "aquaculture" did not specifically recognize the concept of capture-based aquaculture (CBA), and given the extensive practice of CBA, the need for a clear definition for this distinctive activity was recognized.

Bait fish typically refers to smaller pelagic fish species used for bait, for reduction to fish meal, or for direct feeding to carnivorous fish in aquaculture. These species may also be used for human consumption. Typically, fish are classified as bait fish in situations where they are not immediately in demand for human consumption and thus, are considered suitable for use as feeds.

Culture-based fisheries (CBF) are capture fisheries that are maintained by stocking with seed material originating from aquaculture installations (see FAO technical guidelines on Inland Fisheries)[1]. The definition is applicable to both marine and freshwater species.

Ecosystem approach to aquaculture (EAA) is a strategy for the integration of the activity within the wider ecosystem such that it promotes sustainable development, equity and resilience of interlinked social-ecological systems.

Ecosystem approach to fisheries (EAF) strives to balance diverse societal objectives by taking account of the knowledge and uncertainties of biotic, abiotic and human components of ecosystems and their interactions and applying an integrated approach to fisheries within ecologically meaningful boundaries.

---

[1] FAO. 1997. *Inland fisheries*. FAO Technical Guidelines for Responsible Fisheries. No. 6. Rome. 36 pp. (also available at www.fao.org/docrep/003/w6930e/w6930e00.htm).

Enhancement refers to human-induced alterations to natural habitats or application of artificial culture or stocking techniques that are intended to lead to increased abundance. One of the problems with this definition is that it is often very difficult to demonstrate that such "enhancement" has actually occurred and, as such, "enhancement" should not be assumed without due evaluation.

Farming (see Aquaculture).

Illegal, unreported and unregulated (IUU) fishing[2]
Illegal fishing refers to activities conducted by national or foreign vessels in waters under the jurisdiction of a State, without the permission of that State, or in contravention of its laws and regulations; conducted by vessels flying the flag of States that are parties to a relevant regional fisheries management organization but operate in contravention of the conservation and management measures adopted by that organization and by which the States are bound, or relevant provisions of the applicable international law; or in violation of national laws or international obligations, including those undertaken by cooperating States to a relevant regional fisheries management organization.
Unreported fishing refers to fishing activities which have not been reported, or have been misreported, to the relevant national authority, in contravention of national laws and regulations; or undertaken in the area of competence of a relevant regional fisheries management organization which have not been reported or have been misreported, in contravention of the reporting procedures of that organization.
Unregulated fishing refers to fishing activities in the area of application of a relevant regional fisheries management organization that are conducted by vessels without nationality, or by those flying the flag of a State not party to that organization, or by a fishing entity, in a manner that is not consistent with or contravenes the conservation and management measures of that organization; or in areas or for fish stocks in relation to which there are no applicable conservation or management measures and where such fishing activities are conducted in a manner inconsistent with State responsibilities for the conservation of living marine resources under international law.

Low-value fish (see Trash fish).

---

2 Definition from Articles 3.1–3.3 of *FAO International Plan of Action to Prevent, Deter and Eliminate Illegal, Unreported and Unregulated Fishing*. Rome, FAO. 2001. 24 pp.

Overfishing is a generic term used to refer to the state of a stock subject to a level of fishing effort or fishing mortality such that a reduction of effort would, in the medium-term, lead to an increase in the total catch. Often referred to as overexploitation and equated to biological overfishing, it results from a combination of growth overfishing and recruitment overfishing and occurs often together with ecosystem overfishing and economic overfishing.

Precautionary approach involves the application of prudent foresight. Taking account of the uncertainties in fisheries systems and the need to take action with incomplete knowledge, it requires, *inter alia*, consideration of the needs of future generations and avoidance of changes that are not potentially reversible; prior identification of undesirable outcomes and of measures that will avoid them or correct them promptly; that any necessary corrective measures are initiated without delay; that corrective measures should achieve their purpose promptly, on a time scale not exceeding two or three decades; that where the likely impact of the resource use is uncertain, priority should be given to conserving the productive capacity of the resource; that harvesting and processing capacity should be commensurate with estimated sustainable levels of the resource, and that increases in capacity should be further contained when resource productivity is highly uncertain; all fishing activities must have prior management authorization and be subject to periodic review; and an established legal and institutional framework for fishery management, within which management plans that implement the above points, are instituted for each fishery, and appropriate placement of the burden of proof by adhering to the requirements above[3] (see also Appendix 4).

Recruitment overfishing is the rate of fishing above which the recruitment to the exploitable stock becomes significantly reduced. This is characterized by a greatly reduced spawning stock, a decreasing proportion of older fish in the catch, and generally very low recruitment year after year. It may lead to stock collapse if prolonged and combined with poor environmental conditions.

Reference point is a benchmark against which to assess the performance of management in achieving an operational objective, corresponding to a state considered to be desirable (target reference point) or undesirable and requiring immediate action (limit reference point).

---

[3] FAO. 1995. *Code of conduct for responsible fisheries*. Rome. 41 pp. (also available at www.fao.org/docrep/005/v9878e/v9878e00.htm).

Restocking involves the introduction of native or non-native fish or invertebrates reared in hatcheries or captured in and transferred from other areas, where the animals have been produced from hatcheries or grown-out from wild-caught natural resources, to enhance future fish production. Usually, this involves the intention to restore fish reproductive capacity. Restocking can be included in a management programme for restoring habitat quality, recovery of threatened species, or is used independently of a restoration programme. More recently, this activity has been referred to as "culture-based fisheries" (see definition).

Seed is a general term that refers to larvae, post-larvae, fry, fingerlings, juveniles and occasionally adults used for aquaculture grow-out to marketable or consumable size. Wild seed refers to seed fished from the wild (as opposed to produced in a hatchery).[4]

Spawning stock biomass refers to the total weight of all fish (both males and females) in the population which contribute to reproduction. Often conventionally defined as the biomass of all individuals beyond "age at first maturity" or "size at first maturity" i.e. beyond the age or size class in which 50 percent of the individuals are mature.

Stock is a group of individuals in a species occupying a well-defined spatial range independent of other stocks of the same species. Random dispersal and directed migrations due to seasonal or reproductive activity can occur. Such a group can be regarded as an entity for management or assessment purposes, although a stock may or may not be a genetic population. Some species form a single stock while others are composed of several stocks. The impact of fishing on a species cannot be fully determined without knowledge of its stock structure.

Sustainable use refers to the use of components of biological diversity in a way and at a rate that does not lead to the long term decline of biological diversity, thereby maintaining its potential to meet the needs and aspirations of present and future generations.

Trash fish refers to a mixture of small fish caught as bycatch (most typically from trawl fisheries), or as a targeted catch, often used as feed for cultured fish during their grow-out phase. It is referred to as "trash" because formerly

---

4 Modified from Sadovy, Y.J. & Lau, P.F. 2002. Prospects and problems for mariculture in Hong Kong associated with wild-caught seed and feed. *Aquaculture Economics and Management*, 6(3/4): 177–190.

it was taken as bycatch and was not considered to have any economic value. Nowadays, it is valued as feed for fish. It is also valuable in that it comprises the juveniles of many species of direct food importance to humans, or the food for such fish. The term "trash" is, therefore, inappropriate and should be discouraged, to be replaced by the more descriptive term "low-value fish".

Newly developed terms to address specific issues relating to CBA

The following terms have been developed to facilitate discussion on issues relating to capture-based aquaculture.

CBA broodstock fishery refers to the repeated requirement to capture wild, sexually mature individuals (broodstock) to replenish hatchery stocks and/or broodstock populations (i.e. the broodstock is not consistently derived from aquaculture operations). This is distinct from stocks held in captivity that are typically genetically differentiated from wild stocks as a result of their having been bred. Wild broodstock may have to be continually sourced to maintain genetic diversity, replace dead broodstock, or because the production of multiple generations of adult spawners in captivity through the process of full-cycle (hatchery) aquaculture is not biologically feasible or economically viable.

Fattening refers to the placing of wild-captured aquatic animals in captivity and feeding them to increase the size, weight or fat content prior to marketing. Note that "fattening" typically involves a relatively short period of time (i.e. months) but can be longer, as sometimes in the case of tuna, and that "grow-out" typically involves a relatively longer period of time (i.e. many months or even years). However, both activities involve CBA and are used interchangeably in that context. The term "fattening" can also be applied to related short-term grow-out activities, such as crab culture when "empty" crabs that have recently moulted and have not fully grown to fill their new shells are fattened for a few weeks prior to sale.

Hatchery-based aquaculture (HBA) is the farming of aquatic organisms using broodstock produced from farmed stocks through full-cycle aquaculture.

Live (stocked [stocking]) material refers to the live aquatic organisms captured from the wild and placed into an aquaculture operation. This term will encompass all the sizes and life cycle stages – from eggs, larvae and fingerlings, through juveniles to larger even adult fish and broodstock, as well as plants. Live stocked (stocking) material is analogous with the terms seed, seedstock or broodstock that are often used in relation to CBA.

Live storage is the holding of aquatic organisms that are already of market size for the purpose of transport or to await advantageous marketing or price. Maintenance feeding of animals may be provided, as well as other management interventions. However, in live storage, there is no significant or intended incremental increase in weight or size of the animal even though the holding period might be quite extended. Live storage is not considered to be aquaculture.

Appendix 2

CODES OF PRACTICE FOR ALIEN SPECIES

The International Council for the Exploration of the Sea (ICES) and the European Inland Fishery and Aquaculture Advisory Commission (EIFAAC) are two intergovernmental bodies that acknowledge the necessity for international cooperation in order to conserve and use living aquatic resources responsibly. The groups noted the success derived from the growth of marine and freshwater aquaculture and established a set of procedures to be followed in the European and North Atlantic region to address three main challenges from alien species: (i) to reduce the chance of disease transfer from the movement of aquatic species; (ii) to reduce impacts of alien species on native aquatic biodiversity; and (iii) to address the impact that genetically altered stocks may have on related natural populations. These codes and procedures have been endorsed by the Code of Conduct for Responsible Fisheries and have been adopted in principle by other regional bodies of FAO.

The basic code contains the requirements that:

1. the entity moving an exotic species develops a **PROPOSAL** that would include location of facility, planned use, passport information and source of the exotic species;
2. an independent **REVIEW** is made that evaluates the proposal, the impacts and the risks and benefits of the proposed introduction, e.g. pathogens, ecological requirements and/or interactions, genetic concerns, socio-economic benefits and concerns, and local species most affected;
3. **ADVICE** and comments are communicated among the proposers, evaluators and decision-makers, and the independent review **ADVISES** acceptance, refinement, or rejection of the proposal in such a way that all parties understand the basis for any decision or action. Thus, proposals can be refined and the review panel can request additional information on which to make its recommendation;
4. if approval to introduce a species is granted, **QUARANTINE, CONTAINMENT, MONITORING AND REPORTING PROGRAMMES** are implemented; and
5. the **ONGOING PRACTICE** of importing the (formerly) exotic species becomes subject to review and inspection that check the general condition of the shipments, e.g. ensuring that no pathogens are present and that the correct species is being shipped.

These codes are general and can be adapted to specific circumstances and resource availability. However, none of the above requirements should be omitted and nor should the rigour in the application of the requirements be compromised. For example, a regulatory agency may require a proposal to contain a first evaluation of the risks and benefits, and this evaluation would then be forwarded to an independent review or advisory panel; or the advisory panel could make the first evaluation of a proposal. Similarly, States may require quarantine procedures to be described explicitly in the proposal before approval is granted. For additional information, see Bartley, D.M., ed./comp. 2006. *Introduced species in fisheries and aquaculture: information for responsible use and control*. Rome, FAO. (CD-ROM).

Appendix 3

CASE STUDIES OF CBA FISHERIES AND RELATED ACTIVITIES

Box A3.1
Transition from CBA to HBA – work in progress with groupers and sturgeons

Grouper (family: Serranidae) seed produced from hatcheries is reported for about ten species in Southeast Asia, although, few fulfil hatchery-based aquaculture (HBA) entirely and capture-based aquaculture (CBA) is likely to continue into the long term for most species for both economic and practical reasons. Many grouper mariculture operations continue to purchase wild-sourced fish, inclusive of seed for grow-out and of adults for replacement of broodstock on a regular basis. The rate and volume of broodstock are not quantified but are likely to be substantial because broodstock rarely reach second generation and their lives are often shortened by intensive use of chemicals that stimulate breeding. In addition to seed from larval and small juvenile stages, submarket size fish close to the stage of sexual maturation are caught for grow-out to market size. Removal of too much such seed from the wild for CBA could lead to insufficient remaining fish for population replenishment, i.e. recruitment overfishing. Reports from Southeast Asia indicate much reduced numbers of grouper juveniles, although, whether this is due to overfishing of seed and/or adults, or to other factors, is not known.

Understanding the relationships between the number of seed (at different stages), fishing pressure and the status of adult stock is critical for setting quotas of seed collection that do not threaten the stock or population in the long term. It is also important for integrating CBA with other capture pressures on the same population, such as on adults. For example, the red or Hong Kong grouper, *Epinephelus akaara*, of high commercial importance between the 1960s and 1990s in Hong Kong Special Administrative Region, is still an expensive and well-favoured marine food fish throughout its limited geographic range. Although HBA has been possible for the species for more than four decades, only CBA is practised. The species is listed as threatened on the International Union for Conservation of Nature (IUCN) Red List with serious declines in most fishing grounds.

Box A3.1 (Continued)

Sturgeons (family: Acipenseridae) are highly valued for their roe, widely known as caviar. The People's Republic of China has been the largest producer of cultured sturgeons globally since 2000. Between the late 1950s and mid-1970s, artificial reproduction was conducted by collection of mature broodstock and hormone injection in *Acipenser schrenckii*, *A. sinensis* and *A. sabryanus*. By 2002, second generation *A. schrenckii* offspring were being produced by HBA, but for other sturgeons in China, eggs or seed from the wild are still taken and the impact on the stock in the wild of broodstock collection requires assessment and management attention. There is also a need to preserve older, larger sturgeons and safeguard the gene pool of critically endangered species. On the other hand, sturgeons produced in California, United States of America, are captive-raised.

Box A3.2
Transition from CBA to HBA for carp in Bangladesh – can complete transition occur?

The Bangladesh Department of Fisheries successfully produced carp fingerlings in the mid-1960s. It initiated commercial hatchery production in 1975, prior to which aquaculture had been wholly dependent on capture-based aquaculture (CBA) sources. In 1977, the Department of Fisheries established hatcheries as part of a shift from wild sources to hatchery sources of seed supply for grow-out operations. Subsequently, large numbers of hatcheries were established and the collection of wild fertilized eggs and seed declined (also because of habitat loss and degradation). This change from entirely CBA to predominantly hatchery-based aquaculture (HBA) carp culture in the last 20 years in Bangladesh has been assisted by both the public and private sectors. However, dependence on CBA is likely to continue and the transition from CBA to HBA in carp may never be complete as wild broodstock is needed to ensure genetic diversity for HBA operations.

Box A3.3
Culture of yellowtails in Japan – continued CBA despite HBA

Yellowtail (family: Carangidae; jacks) have been cultured in Japan for more than 70 years based on the grow-out of wild caught seed ("mojako") of several species of *Seriola*. Mojako are mainly caught in Japanese waters on drifting seaweed but are sometimes imported. Aquaculture production from capture-based aquaculture (CBA) typically exceeds that produced by wild harvest (i.e. capture fisheries of larger fish). In 1966, in order to conserve the resource, Japan's National Fishery Agency imposed regulations limiting the number of mojako that could be caught annually for the purpose of aquaculture to about 40 million. By 1997, the allocation had dropped to 25 million, such that catch levels are now controlled at less than 2.5 percent of estimated seed stock size (1 billion). Despite such measure, domestic supply has shown significant declines and many mojako are now imported.

Although hatchery-based aquaculture (HBA) is possible, hatchery-produced seed is more expensive than wild-caught seed and farmers prefer to use wild-caught seed over hatchery-produced seed as the latter is not only more expensive, but usually too small for successful rearing. Moreover, hatchery-produced seed has a high percentage of body deformities, and mass seed production has not yet been achieved (mainly because of the difficulty in securing sufficient healthy broodstock). There is an urgent need to address the sustainable use of wild populations and to improve hatchery production to avoid serious depletion of wild yellowtail stocks.

Box A3.4
Tunas – conflicts of interest across multiple fishery sectors

The exploitation of a common resource often creates conflicts between different fishing sectors and fishing related to capture-based aquaculture (CBA) is no exception. In the Mediterranean Sea and adjacent regions, for example, the fishery for bluefin tuna, *Thunnus thynnus*, is among the oldest organized on an industrial scale. The rapid expansion of CBA tuna activities has focused particularly on the purse seine, which is not only a highly efficient fishing method but also the only one that allows for the transfer of live fish to CBA cages. Capture-based tuna aquaculture activities in the Mediterranean have caused friction with local tuna fishers

Box A3.4 (Continued)

using longlines because the activity of tugboats towing live tuna in cages disturbs the traditional longline fisheries, as well as contributing to the reduction of tuna catches in general. In Mexico, an additional conflict of interest is between the CBA farmers and the sardine boat owners. The latter oppose tuna CBA farmers owning and operating sardine boats because they are concerned about loss of control of sardine production and prices. Furthermore, there is competition between the farmers and the sardine-processing plants for limited sardine supply; CBA farmers pay higher prices for fresh sardines than do the frozen sardine packing and fishmeal and fish-oil reduction industries.

Box A3.5
Importance of carp seed fishery to small-scale fishers in Bangladesh

The carp seed fishery in Bangladesh has a long history involving many small-scale activities in the collection, rearing and transportation of carp seeds from river sources to fish farmers for capture-based aquaculture (CBA) grow-out. In the mid-1970s, carp seed, in the form of fertilized eggs or fingerlings, came exclusively from the wild. Part of the catch was reared in the collectors' nursery ponds and part sold to other nursery operators. The wild-carp-seed fishery intensified with the rapid expansion of carp hatcheries in the country and improved fish culture practices in ponds, shifting from a seasonally based livelihood activity conducted by small numbers of specialized fishers to commercial enterprises involving large numbers of poor collectors who otherwise would have remained unemployed for part of the year. These activities included egg and fry collection, broodfish rearing, hatchery operations, transportation, nursery rearing, feed industry and marketing.

The declining availability of natural carp eggs in rivers, increased availability of hatchery-produced carp eggs, and encouragement by the government for farmers to use hatchery-produced seeds have led to a substantial reduction in the number of people involved in the catch of wild carp seed and increased activities around hatchery-produced seed. The wild seed fishery in Bangladesh continues as a small-scale fishery and, although the use of hatchery-produced fry is encouraged, people still catch wild seed while enforcement of legislation to protect wild seed is poorly developed.

Box A3.6
Illegal European eel fisheries

As a result of the high price obtained for glass eels in Asian markets and low availability of glass eels in Asia, many poachers seek glass eels in southern Europe. The future of the European eel population would appear to depend on an intensive battle against poaching, which, in certain areas, involves significant underground activity. Given the difficulties in significantly increasing police surveillance, improved understanding of the fishing and trading networks is important. In France, in order to be allowed to fish and sell eel to a wholesaler or a fishmonger, it is necessary to have a fishing licence. However, a licence holder could, in addition to his/her own harvest, sell on behalf of others. Therefore, there is an element of trust that licence holders are only handling their own catches. A comparison of the numbers of glass eel landed by the professional fishery with the quantity of glass eel actually sold would allow for a better understanding of the extent of black market trading in glass eels. The concern over illegal trading is one reason why the various European Union Member States are interested in developing fishery and trade databases for the European eel for their own eel restoration programmes.

Box A3.7
Illegal mullet seed fisheries in Egypt

The number of mullet fry collected through illegal fishing in Egypt is thought to be considerable, but is not subjected to any form of control. This not only undermines effective management of seed fisheries, but heavily affects the management of other (i.e. non-seed) capture fisheries for the species. The scale of illegal fishing can be very large with the amount of collected fry possibly exceeding the legal harvest. Legal seed fishing is limited to specific sites and to a predetermined number of days each season. Collection sites are selected to avoid disturbing the movements of fry to nursery, feeding and growing grounds. On the other hand, illegal activities take place intensively in protected areas, especially in canals where aggregations of fry migrate from the sea to lakes and coastal lagoons.

Box A3.8

Uncertainties of biomass transfers of Atlantic bluefin tuna – a stock assessment issue

A key to sustainable management planning is stock assessment. For Atlantic bluefin tuna, this requires, among other things, information collected from fishing operations. With the increasing use of capture-based aquaculture (CBA), there are growing uncertainties in officially reported catch data, with the size and age composition of wild fish becoming more difficult to determine with an acceptable degree of precision. In the Mediterranean Sea, during the fishing season, purse seiners capture and transfer schools of tuna at sea from the purse seines into towing cages. Counting of fish trapped within the seine is usually done by divers, while cameras count the fish when they pass from the seine to the towing cage, and average weight is estimated from the dead fish in the seine. Currently, there is insufficient determination of live tuna biometrics, and the resulting uncertainly in the data undermines establishment of effective management measures; the model used to assess the state of the tuna stock should be used with caution because of these increasing uncertainties in the officially reported catches owing to increased CBA. This constitutes a major problem, as the Atlantic bluefin tuna spawning biomass continues to decline while fishing mortality is increasing rapidly, especially for larger fish specimens. Reduction of the uncertainty in biometric statistics data is essential for improving data collection and management of the Atlantic bluefin tuna.

Appendix 4

PRECAUTIONARY APPROACH

The precautionary approach to fisheries management is about being cautious when scientific knowledge is uncertain, and not using the absence of adequate scientific information as a reason to postpone action or failure to take action to avoid serious harm to fish stocks or their ecosystem.

A precautionary approach is, therefore, a set of agreed measures and actions, including future courses of action, that ensures prudent foresight and reduces or avoids risk to the resource, the environment and the people, to the best extent possible, taking into account existing uncertainties and the potential consequences of being wrong (FAO, 1996). FAO technical guidelines on the precautionary approach to fisheries management include precautionary measures for four typical situations: (i) new or developing fisheries; (ii) overutilized fisheries; (iii) fully utilized fisheries; and (iv) traditional or artisanal fisheries (FAO, 1996) (Box A4.1). Some of these will apply to all types of fisheries, whereas others will be useful only in specific situations such as overexploited fisheries. The measures could be included in comprehensive fisheries plans and can also be used in the interim plan for immediate precautionary action until various proposed management plans have been evaluated and approved to replace the interim action.

Box A4.1
Precautionary approach measures

New or developing fisheries

- Always control access to the fishery early, before problems appear. An open-access fishery is not precautionary. Immediately put a conservative cap (or default level) on both fishing capacity and the total fishing mortality rate. This could be achieved by limiting effort or total allowable catch.
- Build in flexibility so that it is feasible to phase vessels out of the fleet, if this becomes necessary. To avoid new investments in fishing capacity, temporarily license vessels from another fishery.
- To limit risks to the resource and the environment, use area closures. Closures provide refuge for fish stocks, protect habitat and provide areas for comparison with fished areas.

Box A4.1 (Continued)

- Establish precautionary, preliminary biological limit reference points (e.g. spawning stock biomass less than 50 percent of the initial biomass) in the planning stage.
  - Encourage fishing in a responsible manner to ensure long term persistence of a productive stock or other parts of the ecosystem.
  - Encourage development of fisheries that are economically viable without long term subsidies.
  - Establish a data collection and reporting system for new fisheries early in their development.
  - Immediately start a research programme on the stock and fisheries, including the response of individual vessels to regulations.
  - Take advantage of any opportunities for setting up experimental situations to generate information on the resources.

Overutilized fisheries

- Immediately limit access to the fishery and put a cap on a further increase in fishing capacity and fishing mortality rate.
- Establish a recovery plan that will rebuild the stock over a specific time period with reasonable certainty.
- Reduce fishing mortality rates long enough to allow rebuilding of the spawning stock.
- When there is a good year class, give priority to using the recruits to rebuild the stock rather than increasing the allowable harvest.
- Reduce fishing capacity to avoid recurrence of overutilization.
- Alternatively, allow vessels to move from an overutilized fishery into another fishery, as long as the pressure from this redeployment does not jeopardize the fishery that the vessels are moving into.
- Do not use artificial propagation as a substitute for the precautionary measures listed above.
- In the management plan, establish biological reference points to define recovery, using measures of stock status, such as spawning stock biomass, spatial distribution, age structure or recruitment.
- For species where it is possible, closely monitor the productivity and total area of required habitat to provide another indicator of when management action is needed.

Box A4.1 (Continued)

Fully utilized fisheries

- Ensure that there are means to effectively keep fishing mortality rate and fishing capacity at the existing level.
- There are many "early-warning signs" that a stock is becoming overutilized (e.g. age structure of the spawners shifting to an unusually high proportion of young fish, shrinking spatial distribution of the stock or species composition in the catch). These warning signs should trigger investigative action according to prespecified procedures while interim management actions are taken, as noted below.
- When precautionary or limit reference points are approached closely, prespecified measures should be taken immediately to ensure that they will not be exceeded.
  - If limit reference points are exceeded, recovery plans should be implemented immediately to restore the stock. The recommendations for overutilized stocks described above should then be implemented.
  - To prevent excessive reduction of the reproductive capacity of a population, avoid harvesting immature fish unless there is strong protection of the spawning stock.

Traditional or artisanal fisheries

- Keep some areas closed to fishing in order to limit risks to the resource and the environment.
- Delegate some of the decision-making, especially area closures and entry limitations, to local communities or cooperatives.
- Ensure that fishing pressure from other (e.g. industrial) segments of the fishery does not deplete the resources to the point where severe corrective action is needed.
- Investigate the factors that influence the behaviour of harvesters to develop approaches that can control fishing intensity.

Source: FAO (1996).